British Association for the Advancement of Science

Handbook of Canada

British Association for the Advancement of Science

British Association for the Advancement of Science

Handbook of Canada
British Association for the Advancement of Science

ISBN/EAN: 9783337208868

Printed in Europe, USA, Canada, Australia, Japan

Cover: Foto ©berggeist007 / pixelio.de

More available books at **www.hansebooks.com**

HANDBOOK

OF

CANADA

BRITISH ASSOCIATION

FOR THE

ADVANCEMENT OF SCIENCE

TORONTO MEETING 1897

HANDBOOK

OF

CANADA

PUBLISHED BY THE

PUBLICATION COMMITTEE OF THE LOCAL EXECUTIVE

TORONTO

1897

List of Charts and Diagrams.

AT BEGINNING OF VOLUME.

AT END OF VOLUME.

PREFACE.

THIS Handbook has been designed to give for the use of scientific visitors to Canada a compact and systematic account of those features of the Dominion in which they might be presumed to have a special interest. The essential condition of the volume was that it should not be bulky, and this has necessarily imposed severe limitations alike as regards amplitude of treatment and as regards choice of topics. The Publication Committee are indebted to the authors for their hearty and gratuitous co-operation in the production of the volume. The map which is placed at the beginning is entitled to be regarded as the most recent map of Canada. It has been specially prepared for this volume under the direction of Dr. G. M. Dawson, and it embodies the most recent nomenclature, and the results of the most recent exploration in the Canadian portion of the Arctic region. The meteorological charts have also been specially prepared for this volume under the direction of Mr. R. F. Stupart. The volume has been edited by Professor Ramsay Wright and Professor James Mavor.

PART I

THE GEOGRAPHY, GEOLOGY, CLIMATOLOGY
AND BIOLOGY OF CANADA.

PART I.

THE GEOGRAPHY, GEOLOGY, CLIMATOLOGY AND BIOLOGY OF CANADA.

CHAPTER I.

Physical Geography and Geology.

By George M. Dawson, C.M.G., F.R.S., Director of the Geological Survey of Canada.

THE Dominion of Canada embraces the northern half of the continent of North America, with its adjacent islands, including those of the Arctic Ocean between the 141st meridian and Greenland, but exclusive of Alaska, in the extreme north-west, the island of Newfoundland, which still remains a separate British colony, and the small islands of St. Pierre and Miquelon, retained by France. In order, therefore, to include the sister colony of Newfoundland, the general name, British North America must still be employed. The total area of Canada is estimated at about 3,574,-980 square miles, of which the Arctic islands to the north of the continental land make about 309,000 square miles. Thus, with Newfoundland (42,000 square miles), the aggregate area of British North America is 3,616,980 square miles. This area is somewhat larger than the United States (including Alaska), and not much less than all Europe.

The form of the North American continent may be described as that of an isosceles triangle, of which the narrower part, pointing to the south, constitutes Mexico, a wide central belt, the United States, while the broader base is the Dominion of Canada. The northern base-line of the continental land lies approximately on the seventieth parallel of north latitude, but is broken into by the geographically important sea named Hudson Bay, 800 miles from north to south, and some 600 miles in width. Historically, this great indentation of the northern land has been notable in connection with the exploration and trade of British North America, and it promises in the future, when modern means of communication have been adapted to the circumstances, again to become important. But the ruling physical features upon which the existence of Canada as a country depends, and about which its history has grown up, are, the proximity of the northeastern part of the continent to Europe, and the existence of a great waterway, the River St. Lawrence, running to the very centre of the continent, and expanding there into the group of inland seas generally spoken of as the Great Lakes. The first of these made the landfall of Cabot in 1497 possible, the second led Cartier (in 1535) to the sites now occupied by the cities of Quebec and Montreal, and opened a route of exploration and commerce to subsequent explorers and traders of France, by which they overran much of the central and western country of North America before the colonists of New England, further to the south, achieved a way across the Appalachian highlands which there barred their progress.

The course of the St. Lawrence and the position of the Great Lakes may be regarded as being determined by the southern outline of the Laurentian plateau, composed of ancient crystalline rocks, which in U-shaped form surrounds the central depression of Hudson Bay. Spreading widely in the Labrador peninsula, this tract of relatively high and generally rocky land, runs with narrower dimensions round the southern extremity of the bay, and thence is continued north-westward to the Arctic Ocean. With one important exception, that of the Ontario peninsula, which juts far to the south, the great river and its reservoirs lie along the southern edge of these highlands, while the Winnipeg system of lakes, Athabasca, Great Slave and Great Bear Lakes, occupy a very similar position on the outer rim of the north-western extension of the plateau.

Suess has named this plateau, composed of Archæan rocks, the "Canadian shield." In respect to form, it is the most important geological feature of the continent, for, under whatever name, it must be considered its nuclear tract or "protaxis." About it the later sediments, from Cambrian times to the present day, have been spread, their material being largely supplied by its waste, and against it or parallel to its main outlines, successive movements of the earth's crust have forced up the newer and now higher ranges of the Appalachian and the Cordilleran systems, while, so far as we are able now to determine, the ancient shield itself has remained comparatively unaltered and fixed. Following the trends of the south-east and south-west sides of the Laurentian highlands respectively, the Appalachian mountains and those comprised in the western Cordillera converge to the south, embracing between them, to the south of the Great Lakes, the central plain of the continent, based upon comparatively undisturbed and nearly horizontal strata, which it may reasonably be conjectured repose, at no very great depth below the surface, upon a now concealed southern extension of the original platform. But there is an important want of symmetry in these main orographic features, for, between the western edge of the Laurentian highlands and the Cordilleran ranges, a wide tract of the central plain runs northward to the Arctic Ocean, constituting in its southern part, the great arable inland country of western Canada.

From a physiographical and geological standpoint, the Canadian part of the continent may very naturally be regarded as composed of two great divisions—an eastern and a western; the line between these beginning at the south near Winnipeg, on the central longitude, and running thence along the outer edge of the Laurentian plateau north-westward to the Arctic Ocean. To the east of this line, while the surface is generally broken and irregular, the relief is nearly everywhere comparatively low. The rocks are almost altogether referable to the Palæozoic systems or to systems older than these, and there is little evidence of important change during the later geological periods, beyond such as is incident to the gradual wearing away and denudation of very ancient highlands and mountain systems.

To the west, the Mesozoic and Tertiary systems become important. The entire spread of the great plains is floored by such rocks, and they occupy also a large part of the western Cordilleran belt, although there mingled with important areas of much older rocks. At a late date, geologically, the Pacific Ocean has spread over nearly all this part of the continent. The vast orogenic changes which have resulted in excluding the sea from the region are comparatively recent; many of the mountain ranges of the Cordillera are rugged, new and lofty, and the processes of denudation are still going on very rapidly, with rivers and streams flowing at high grades, and very far from that passive condition found where the drainage system has approximately reached the base-level of erosion.

A two-fold division of the northern part of the continent, of the kind above indicated, although based upon fundamental facts in its geological history, is, however, much too general for the purposes of description of its several regions as they present themselves to us to-day. The characteristically mountainous region of the Cordillera, and that of the great central plain, must be separately treated; while the great eastern division is not only larger than that of the west, but, being in its southern portion the home of the greater part of the population, requires closer consideration. The boundaries of the several provinces, resulting from circumstances of a more or less political kind, do not always correspond with the natural features, and cannot therefore be adopted as the best for purposes of geographical and geological description. Relying chiefly upon the physical and geological facts, we may therefore further subdivide the continental part of Canada as follows :—

(1.) *The Acadian Region*, including the Maritime Provinces of the Atlantic, with the similarly characterized south-eastern part of the Province of Quebec, bounded by a line running from the Strait of Belle Isle to the City of Quebec, and thence to Lake Champlain.

(2.) *The Lowlands of the St. Lawrence Valley*, extending, with an irregular width, from the City of Quebec to Lake Huron, and including the Ontario peninsula.

(3.) *The Laurentian Plateau*, which, notwithstanding its vast area, may, from a physiographical point of view, be regarded as a unit.

(4.) *The Interior Continental Plain*, running with narrowing dimensions from the 49th parallel to the Arctic Ocean, and embracing part of Manitoba and the North-West Territory.

(5.) *The Cordillera*, or great mountain belt of the west, including the greater part of British Columbia and the whole of the Yukon district.

Before entering upon the description of the several regions thus defined, something may be said of the drainage system of the northern part of the continent as a whole. Only the most general notice can be given to the rivers and lakes of Canada in a review so short as this must necessarily be, but no feature of the country is more important, whether historically or geographically, than the great length and volume of its principal watercourses, and the manner in which these interlock and penetrate almost every part of its area. Besides the St. Lawrence, with its drainage basin of 530,000 square miles, to which allusion has already been made, there are three more rivers of the first class, of which the watersheds are wholly or in great part included by Canada. These are the Nelson, the Mackenzie and the Yukon. The first-named reaches Hudson Bay, bringing with it the waters of the Saskatchewan and other large and long rivers which drain a vast region in the centre of the continent. Its basin is estimated at 367,000 square miles. The Mackenzie, flowing into the Arctic Ocean, drains not only most of the northern part of the interior plain of the continent, but also considerable portions both of the Rocky Mountain region and the Laurentian plateau, with a basin of about 677,000 square miles. Next to the St. Lawrence, it is the longest river of Canada, being not less than 1,800 miles from its source to its mouth. The Yukon, discharging into the northern part of Behring Sea, unwaters a great tract of the northern part of the Cordilleran region comprised in Canada, besides flowing across the whole width of Alaska.

It is only by contrast with these greatest rivers that many more are relegated to a second or third rank, as an examination of the map will show. It will also be apparent that much the larger part of the country lies on the northern slope of the continent, regarded as a whole, and that the remainder is divided between the Atlantic and Pacific sides, but an inconsiderable region being tributary to the southward-flowing system of the Missouri and its branches.

It may be useful, in this connection, to state the heights of a few of the larger lakes, as ruling features in physical geography. The Great Lakes, although

they stand at four levels, in reality occupy only two
distinct stages, separated by the Niagara Falls. Below
this cataract is Lake Ontario, 247 feet, above it Lake
Erie, 573 feet, Lakes Huron and Michigan, 581 feet,
and Lake Superior, 602 feet.

Further to the west and north-west are, Lake of the
Woods, 1,062 feet, Lake Winnipeg, 710 feet, Lakes
Manitoba and Winnipegosis, 810 and 830 feet, respectively, Athabasca Lake, 690 feet, Great Slave Lake,
about 520 feet, and Great Bear Lake, about 340 feet.
Each of these lakes marks the lowest level of large
tracts of adjacent land.

The Acadian Region, regarded as a whole, forms
the north-eastward continuation of the Appalachian
mountain system. This mountain system, beginning
not far from the Gulf of Mexico, gives form to the
eastern coast of the United States, from which it is
never far distant. In Vermont and New Hampshire, it is represented by the Green and White Mountains, and its main line runs on, though with much
decreased elevation, through the south-eastern part of
the Province of Quebec, under the name of the Notre-
Dame Mountains. Not far below the City of Quebec
it approaches the St. Lawrence, and thence continues
parallel with that river and its great estuary, all the
way to Gaspé, on the open gulf. In the Gaspé peninsula it is known as the Shickshock Mountains. Considerable parts of these mountains rise well above 3,000
feet, but the Notre-Dame range seldom exceeds
1,000 or 1,500 feet, and its elevations resemble rolling
and broken hills and ridges, rather than mountains
properly so called. The whole length of this main
continuation of the Appalachian system in Canada is
about 500 miles.

Subordinate and less continuous elevations, nearly
parallel to the main range thus outlined, occur in New
Brunswick, chiefly along two lines, one of which strikes
the Baie des Chaleurs below its head, the other, some-

what divergent in direction to the eastward, bordering
the southern shore of the province along the Bay of
Fundy.   The rocks of all these ranges are in the main
older than those of the Carboniferous system, and between the two last-mentioned ridges, in New Brunswick, lies a broad triangular area of nearly horizontal
beds referable to the Carboniferous formation. Besides
this large level area, there are many others of lesser size
and numerous large valleys, comprised in what has
been designated the Acadian region, in Quebec and
New Brunswick.   These often afford excellent arable
land, or support valuable forests.   The character of
the soil varies greatly, chiefly in conformity with that
of the subjacent rocks ; but it has also been considerably
affected, as in almost all parts of Canada, by the nature
and amount of the deposits due to the glacial period.

In the general and inclusive sense in which the
Though lying at some distance to the south-eastward
of the main line, the peninsula of Nova Scotia may
best be regarded as a member of the Appalachian system of uplifts, with which it is parallel.   Its elevation
nowhere exceeds 1,200 feet, and is in general very
much less.   A broad range of broken hills and uplands
extends along the Atlantic coast of the province and
into the island of Cape Breton.   The Cobequid Hills
are next in importance to this, running to the north
of that arm of the Bay of Fundy known as Minas
Basin, and joining the last at an angle, near the middle
of the province.   The Atlantic coast range is chiefly
composed of old Cambrian rocks and granites, with
little land of agricultural value, but rich in gold-
bearing veins, which occur in all parts of its length.
With the disturbed area of the Cobequids, rocks as
new as the Devonian are largely involved.   The best
arable lands of Nova Scotia are situated towards the
Bay of Fundy, and along the northern side of the
peninsula generally.   Here also, and in Cape Breton,
the important coal-fields occur.

In the general and inclusive sense in which the

term "Acadian region" has been employed, this has
thus a width of about 350 miles, between the outer
coasts of Nova Scotia and the St. Lawrence estuary.
Followed to the south-eastward, this belt of country
embraces the New England States and part of New
York, all with very similar physical characters. In
the opposite direction it is interrupted by the Gulf of
St. Lawrence, but reappears in the great island of Newfoundland, still preserving most of its characteristic
features, although somewhat modified in appearances
by differing climatic conditions. Throughout this
region, including Newfoundland, the geological structure is alike, the formations represented are nearly the
same, and, both in composition and from a palæonto-
logical standpoint, they often resemble those of the
opposite side of the Atlantic more closely than they do
those of other parts of North America.

Before speaking of the geological features of the
Acadian region in Canada, two exceptional areas
within its limits may be referred to. In the semi-
circular bay formed by the coasts of New Brunswick
and Nova Scotia, separated from the mainland by
Northumberland Strait, is Prince Edward Island, a
province by itself, although not much more than 2,000
square miles in area. This lies opposite the Carboniferous and Permo-Carboniferous lowlands of eastern
New Brunswick and northern Nova Scotia, and consists entirely of undisturbed and unaltered Permo-
Carboniferous and Triassic red sandstones and shales.
Characteristic fossils of both formations have been
found, but the circumstances render it difficult to draw
a line between them. The surface of the island is for
the most part fertile and highly cultivated, nowhere
exceeding 500 feet above the level of the sea.

The second area of an exceptional character, is that
of the island of Anticosti, lying in the wide estuary of
the St. Lawrence, and 140 miles in length. This again
consists of nearly flat-lying rocks, chiefly of the

Silurian with some of Cambro-Silurian or Ordovician age (Hudson River) along its northern side. The island evidently represents part of a submerged and undisturbed Cambro-Silurian and Silurian tract of the northern part of the Gulf of St. Lawrence, the rocks of which differ in some respects from their representatives further to the west. The island is generally wooded, and is at present very scantily inhabited.

The geological scale is well represented in Nova Scotia and New Brunswick, from the Archæan to the Triassic, but thereafter ensues a long gap, during which no deposits appear to have been formed, probably because the area in question then existed as land, exposed to denuding agencies alone. Closing this unrepresented lapse of time we find only the clays and sands referable to the glacial period, with still more recent deposits, such as those of the fertile marsh lands of the Bay of Fundy.

With the exception of the flat-lying tracts between the several axes of elevation the wide Carboniferous area in New Brunswick and marginal developments of rocks of the same age along the northern coast of Nova Scotia and in Cape Breton, the region must be considered as one of exceptional geological disturbance and complexity, which, notwithstanding the large amount of investigation it has received, is still but imperfectly understood. One cause of difficulty lies in the existence, at several horizons, of thick masses of strata composed of ancient volcanic materials, generally without organic remains. This is a character common to the rock-formations of most parts of the Appalachian region of North America, which has been at many times the theatre of great volcanic activity. The occurrence of similar rocks in the south-eastern part of the Province of Quebec has been the cause of much of the uncertainty attaching to the understanding of the "Quebec group" there.

The region is not a typical one for the Archæan

rocks, but areas of crystalline schists referable to this time occur. Most of these have so far been mapped simply as pre-Cambrian, for no separation into groups has yet been effected. A large tract of the kind occupies the northern part of Cape Breton. In the southern highlands of New Brunswick, particularly in the vicinity of the city of St. John, the Archæan is better characterized and bears a resemblance, almost amounting to identity, with the typical developments of these rocks in Quebec and Ontario. A Lower Laurentian series comparable with the " Fundamental gneiss " is found, with an upper series composed of crystalline limestones, quartzites, etc., like the Grenville series. Newer than these is a mass of strata composed chiefly of volcanic materials, breccias or agglomerates, greenstones and felsites, which is referred to the Huronian system. Further areas of the same kind occur in central New Brunswick.

Forming the backbone of the peninsula of Nova Scotia and bordering its whole Atlantic coast, is a belt characterized by granitic masses and old stratified rocks assigned to the Lower Cambrian. This belt is widest in the south-west part of the province and narrows gradually in the opposite direction. The granites date from about the Devonian period, and it was probably in connection with their intrusion that the bedded rocks were thrown into the great series of parallel, sharp flexures, in which they now lie and which bear so intimate a relation to the systems of gold-bearing veins. The disturbances incident to this period are, in fact, those which have chiefly given form to the Maritime Provinces. All the older rocks are involved in them, while those of the Carboniferous remain comparatively unaffected.

These Cambrian rocks consist of a lower quartzite series and an upper clay-slate or argillite series, the latter generally dark in colour. It must be added that no really characteristic fossils have yet been obtained from these gold-bearing rocks. Distinctive Cambrian faunas have been found only in a few places in Cape Breton.

In the vicinity of St. John, New Brunswick, a remarkably interesting and complete section of Cambrian rocks, well characterized by fossils, occurs. This may now be regarded as typical for the eastern part of Canada and Mr. G. F. Matthew, to whom its elaboration is chiefly due, remarks particularly the resemblance of the fauna to that found in rocks of the same age around the Baltic sea, rather than to that of the interior of North America. In time, the series runs from the Etcheminian (older than the *Olenellus* zone) to the highest Cambrian.

Rocks of Cambro-Silurian (Ordovician) age have also been determined in the locality just referred to, and sparingly in other parts of New Brunswick. In Nova Scotia, the only fossils of the kind come from a single place in Cape Breton, but considerable areas supposed to be of this age, chiefly composed of volcanic materials, occur in other parts of the province.

Silurian (Upper Silurian) rocks, are widely spread in northern New Brunswick, and in adjacent portions of Quebec, occupying the greater part of the area which drains to the Baie des Chaleurs. They recur in the southern part of New Brunswick and in the northern part of Nova Scotia, and although comprising limestones, sandstones and shales, are often greatly intermixed with contemporaneous volcanic materials, indicating, it will be observed, a third important volcanic interlude in the history of this part of the continent. In Nova Scotia, important bedded iron ores (hæmatite) appear in this series. The geological horizons represented, as compared with those of the New York scale, range from the Clinton to the Lower Helderberg.

In the rocks of Devonian age, occurring in that portion of the Appalachian region covered by Nova Scotia, New Brunswick, the Gaspé peninsula of Quebec, and a part of the adjacent State of Maine, a remarkably full representation of the flora of that ancient period is found, from which more than 125 species of land plants have been catalogued by Sir J. Wm. Dawson; while at Scaumenac, on the Baie des Chaleurs, rocks of this age have yielded many interesting fish remains, investigated by Mr. J. F. Whiteaves, which compare closely with those of the Old Red Sandstone. Further west, in beds of the same period in Ontario and New York, the fossils are chiefly marine molluscs, with comparatively little evidence of the existence of adjacent land. The lower part of the Devonian of the Maritime Provinces holds a similar fauna, and contains in Nova Scotia bedded fossiliferous iron ores. The geographical extent of the Devonian rocks in Nova Scotia and New Brunswick is comparatively limited, but in the Gaspé peninsula of Quebec it becomes somewhat important.

The Carboniferous system, both from its extent and because of its economic value, must be considered as one of the most important features of Nova Scotia and New Brunswick, although there is reason to believe that much larger tracts of this formation still lie beneath the waters of the Gulf of St. Lawrence and the Atlantic. Its total thickness is, in some parts of Nova Scotia, estimated at 16,000 feet, but it is very irregular in this respect and over the greater part of New Brunswick is comparatively thin. At the Joggins, on the north arm of the Bay of Fundy, is a remarkable continuous section showing 14,570 feet of strata, including more than seventy seams of coal, which has been made the subject of investigation by Logan, Lyell, and Sir J. Wm. Dawson. From beds in this section numerous specimens of a land-inhabiting reptilian fauna have been described. The flora of the period is well represented in many places, particularly in Nova Scotia, and includes that of several distinct stages, beginning with the Horton group at the base (comparable with the " calciferous sandstone " of Scotland and the lower part of the sub-Carboniferous of Ohio, Tennessee and West Virginia,) and at the top containing so many forms referable to the Permian, that the name Permo-Carboniferous has been applied to this part of the section.

Several local unconformities have been determined in different parts of this great succession of beds. With the marine limestones, important deposits of gypsum are found. The workable coal-seams occur in what is called the Middle Carboniferous, and some of these, in the Pictou district, are of unusual thickness. Coal mining is actively in progress in Cumberland and Pictou counties and in Cape Breton, the total annual output being between two and three million tons. In New Brunswick, the productive area for coal appears to be small, and the seams so far found are of inconsiderable thickness.

To complete this very brief review of the geology of what has been called, broadly, the Acadian region of eastern Canada, it now only remains to add a few words concerning that main line of uplift and disturbance, the course of which was first traced through the Province of Quebec, from the vicinity of Lake Champlain to Gaspé. This structurally complicated belt of country has been the subject of much controversy, and possesses now a literature of its own. It is bounded to the north-westward by an important dislocation or break, known as the St. Lawrence and Champlain fault, which may be traced from Lake Champlain to Quebec City, and thence follows the estuary of the St. Lawrence, probably running to the south of Anticosti. To the west of this line are the flat-lying Cambro-Silurian strata of the St. Lawrence plain, chiefly limestones, and doubtless resting upon a strong shelf of the Laurentian nucleus at no great depth. Against this stable edge, the eastern strata have been folded, faulted and ridged up by the forces which produced the Appalachian range. Were this all, a careful study of the beds on the two sides of the line would readily show their identity ; but, it appears, that previous to the great epoch of disturbance the original physical conditions themselves differed. To the west, a sheltered sea came into existence about the close of the Cambrian period, in which Cambro-Silurian strata, in large part limestones, were laid down. To the east, sedimentation began much earlier, and the circumstances of deposition were different and more varied. Even the animal life present in the two districts was largely dissimilar at the same period, probably as a result of different temperatures in the sea water. Thus it was not until much study and thought had been given to the problem that Logan was enabled to affirm the equivalency of a great part of the strata on the two sides of the St. Lawrence and Champlain fault. To those on the east, differing in composition and fauna from the rocks of the typical New York section, he applied the name "Quebec group." Regarded as a local name attached to the Atlantic type of the lower members of the Ordovician, and distinguishing these from rocks of the same date deposited upon the Continental plateau, this term may still be employed with advantage, expressing as it does a most important fact. Subsequent investigations have shown, however, that in the ridging up of this part of the Appalachian region, not only are some very old Cambrian rocks brought to the surface, but considerable areas of crystalline schists which are evidently pre-Cambrian, and very possibly referable to the Huronian. These were originally included in the "Quebec Group," under mistaken ideas of metamorphism, but their elimination from it, now rendered possible, does not detract from the merit of the original discovery, nor does it impair the value of the term "Quebec group," as a descriptive one, if properly limited.

2

In this folded and disturbed region of south-eastern
Quebec, copper ores, asbestos and chromic iron are
among the more important minerals of value.   Silurian
rocks and some of Devonian age rest in places uncon-
formably upon the older corrugations.

Some facts respecting the glacial deposits of the
Appalachian region are given on a later page, with
general statements relating to this period in eastern
Canada.

Lowlands of the St. Lawrence Valley.   The tract of
country which it is found convenient to include under
this name, comprises but a small part of the hydrographic basin of the great river, which in all is about
530,000 square miles in extent.   Nor is it altogether
uninterrupted, although clearly enough defined in a
general way by the edge of the Laurentian plateau on
the north, the Appalachian highlands to the south-
east, and on the south, further west, by the line of the
St. Lawrence River and the lower members of the system of the Great Lakes.   It may be described as extending from a short distance below the city of Quebec to
Lake Huron, with a length of over 600 miles and an
area of more than 35,000 square miles, all of which
may be regarded as fertile arable land—the greatest
connected spread of such land in eastern Canada.

These lowlands are based upon nearly horizontal
strata, ranging in age from the latest Cambrian (Potsdam sandstone) to the Devonian.   On a geological map
its limits are readily observable; but in order to understand its character, it is necessary to consider it
somewhat more clearly, and under such scrutiny it is
found to break up naturally into three parts.   The
first of these is divided between the Provinces of Quebec and Ontario, running west along the St. Lawrence
and its great tributary the Ottawa somewhat beyond
the 76th meridian, or to a north-and-south line drawn
about twenty-five miles west of the city of Ottawa.
It is here interrupted by a projecting, but not bold,

spur of the Laurentian plateau, which crosses the St.
Lawrence at the lower end of Lake Ontario, forming
there the Thousand Islands, and runs southward to
join the large Archæan tract of the Adirondacks in the
State of New York. This Eastern division, with an
area of 11,400 square miles, constitutes what may be
called the St. Lawrence plain proper, parts of which
were among the first of those occupied by the early
French settlers. Much of its surface is almost absolutely level, and it nowhere exceeds a few hundred
feet in elevation above the sea, although a few bold
trappean hills stand out in an irregular line, with
heights of 500 to 1,800 feet. Mount Royal, at Montreal, is one of these, and from it all the others are in
sight, while the Laurentian highlands may also be seen
thirty miles to the north, and to the southward the
Green Mountains and Adirondacks, forming the boundary of the plain in that direction, are apparent on a
clear day.

Beyond the projecting spur of ancient crystalline
rocks above referred to, from the lower end of Lake
Ontario, near Kingston, to Georgian Bay of Lake Huron, the southern edge of the Laurentian plateau runs
nearly due west, with a slightly sinuous line, for 200
miles. Between this edge and Lake Ontario on the
south, lies a second great tract of plain, the lowest
parts of which may be considered as level with Lake
Ontario (247 feet) but of which no part exceeds 1,000
feet above the sea. This plain is naturally bounded to
the south and west by the rather bold escarpment of
the Niagara limestone, which, after giving rise to the
Falls of Niagara between Lakes Ontario and Erie, runs
across this part of the Province of Ontario to Lake
Huron, forming there a long projecting point and con-
tinuing still further west, in the chain of the Manitoulin Islands. The city of Hamilton lies close under a
part of this escarpment. The area of this second tract
of plain is about 9,700 square miles. It is scarcely

more varied in its surface than that to the eastward,
and most of its extent is a fertile farming country.

The third and last subdivision of the lowlands of
the St. Lawrence valley, is an area of triangular form
included between the Niagara escarpment and Lakes
Erie and Huron. This constitutes what is generally
known as the Ontario peninsula, and its south-western extremity touches the 42nd parallel, the latitude
of Rome. The area of the Ontario peninsula is 14,200
square miles, and both in soil and climate it is singularly favored. Grapes, peaches and Indian corn, or
maize, are staple crops in many districts. To the
north, some parts of this tract are high and bold, but
most of its surface varies from 500 to 1,000 feet above
the sea.

The geological features of the lowlands of the St.
Lawrence valley are comparatively simple. The rocks
flooring the region lie either horizontally or at very low
angles of inclination upon the spreading base of the
Archaean mass to the northward, the crystalline rocks
of which have frequently been met with in deep borings. The formations represented correspond closely
with those of the New York section, which had been
rendered typical by the researches of James Hall and
his colleagues, before any definite examination of
the geology of Canada was begun. Hall's nomenclature has been adopted for the series, which, beginning with the Potsdam sandstones, continues upward
without any marked break, to the Chemung or Later
Devonian.

In the first or eastern subdivision of this region, the
Potsdam sandstone, although strictly speaking referable to the Upper Cambrian, physically considered is
really the basal arenaceous and conglomeritic member
of the Cambro-Silurian (Ordovician) series which follows. The several members of the Cambro-Silurian
occupy almost the entire surface, diversified merely by
a few light structural undulations, which to the south

and east of Lake St. Peter result in the introduction
of some higher beds that are referred, although with
some doubt, to the Silurian. Fossils referable to this
period are also found associated with a small patch of
volcanic breccia on St. Helen's Island, opposite Montreal. This is probably connected with the adjacent
igneous mass of Mount Royal, and by it the age of
this mass, as well as that of the other similar prominent points of the same character in the vicinity, is
pretty definitely fixed as latest Silurian or Lower
Devonian. It would in fact appear that these igneous
masses represent the necks of ancient volcanic eruptions, which have since been subjected to prolonged
denudation. In their vicinity the sedimentary rocks
are shattered and disturbed and traversed in all directions by trap dykes. Basic rocks (theralite, etc.,) together with nepheline-syenites are characteristic.

Passing to the second, or central subdivision, to the
west of Kingston, the Cambro-Silurian formations just
referred to are found to be repeated, in ascending order, along the north shore of Lake Ontario, with very
similar characters and equally undisturbed. The
Trenton limestone occupies the greatest area, extending in a wide belt to Georgian Bay of Lake Huron.
Above this lie the Utica shales, and over these the
Hudson River formation, upon which Toronto is situated. This is the highest member of the Cambro-Silurian, but the plain also overlaps the lower members
of the succeeding Silurian system irregularly, finding
its natural boundary, from a physical point of view,
only at the massive outcrop of the Niagara limestone.

The course of the escarpment produced by this outcrop has already been traced; above it and to the
south-west, lies the higher plain generally known as
the Ontario peninsula, constituting the third subdivision of the St. Lawrence lowlands. More than
half of the area of this peninsula is occupied by Devonian rocks, which succeed the Silurian regularly in

ascending order, the highest beds being met with in
the extreme south-west of Ontario, beyond which, they
are soon followed by the Carboniferous basin of the
Michigan peninsula. The Silurian and Devonian
strata are affected only by slight and low undulations,
but these are important in connection with the exploitation of the oil and gas of the region.

In the two eastern subdivisions of the lowlands of
the St. Lawrence valley, with the exception of structural materials, such as stone, lime and clay, minerals
of economic value are scarcely found; but in the third
or westernmost subdivision, in addition to these, gypsum, salt, petroleum and natural gas have become important products. The gypsum and salt are derived
from the Onondaga formation of the Silurian. The
salt is obtained in the form of brine, from deep wells,
but beds of rock-salt are known to occur at considerable depths. At Goderich, two beds of very pure salt
have been proved by boring to exist at a depth slightly exceeding 1,000 feet, with an aggregate thickness of
over 60 feet. Petroleum is chiefly derived from the
Corniferous limestone of the Devonian, and natural gas
is obtained from several horizons both in the Devonian
and Silurian.

The Laurentian Plateau. The great region thus
named, composed of very ancient crystalline rocks, has
an area of over 2,000,000 square miles, or more than
one-half that of the entire Dominion of Canada. In a
horse-shoe-like form, open to the north, it surrounds
three sides of the comparatively shallow sea known as
Hudson Bay. Its southern part is divided between
the Provinces of Quebec and Ontario, its eastern side
expanding into the Labrador peninsula, while the
western runs, with narrower dimensions, to the
Arctic Sea, west of the great bay.

In geographical extent it is thus very important,
although somewhat monotonous in its physical and
geological features. It contributes little to the fertile

areas of the country in proportion to its size, but in
the aggregate, comprises a considerable amount of land
which is either cultivated or susceptible of cultivation.
Elsewhere, in its southern parts, it carries forests of great
value, and its mineral resources are already known in
some places to be very important. It constitutes,
moreover, a gathering ground for many large and
almost innumerable small rivers and streams, which, in
the sources of power they offer in their descent to the
lower adjacent levels, are likely to prove, in the near
future, of greater and more permanent value to the
industries of the country than an extensive coal field.
Particularly notable from this point of view is the
long series of available water powers which runs from
the Strait of Belle Isle nearly to the head of Lake
Superior, coincident with the southern border of the
plateau.

Although it is appropriate to describe this region as
a plateau or table-land, such terms, it must be understood, are applicable only in a very general way. Its
average elevation of about 1,500 feet is notably greater
than that of the adjacent lands, and is maintained with
considerable regularity, but its surface is nearly everywhere hummocky or undulating. Away from its
borders, the streams draining it are, as a rule, extremely
irregular and tortuous, flowing from lake to lake in
almost every direction; but assuming more direct and
rapid courses in deeply cut valleys as they eventually
leave it. Many of the surface features are of very
great antiquity, Mr. Low's observations in Labrador
going to show that the larger valleys there existed
much in their present form before the Cambrian period.

The average height of the central parts of the
Labrador peninsula is about 1,700 feet, and the most
of its drainage is divided between Hudson Bay, Ungava
Bay and the Atlantic coast, the main watershed lying
not very far to the north of the St. Lawrence estuary
and gulf. Along the Atlantic coast, to the north of

Hamilton Inlet, the region assumes a really mountainous character, numerous elevations attaining 3,000 feet,
and some as much as 5,000 or 6,000 feet.   These are
the highest known points connected with any part of
the Laurentian region, and are quite exceptional in
character.   The same high range is continued to the
north of Hudson Strait in the western part of Baffin
Land.

To the south of Hudson Bay, the watershed is, at
least in one part, as low as 1,000 feet.   East of Lake
Winnipeg, the Nelson and Churchill Rivers cross the
Laurentian plateau, in a wide depression, to reach
Hudson Bay.   Still further north, this part of the
plateau has a height of about 1,100 feet above the
sea.

Generally speaking, the surface of the plateau is
barren and rocky, though often with wide swampy
tracts towards the height-of-land.   To the south-west
of Hudson Bay it is overlapped by an important spread
of Silurian and Devonian rocks, over which, and the
adjacent parts of the crystalline rocks, is rather uniformly spread a mantle of alluvial deposits, affording a good
soil, which in some places may eventually be of value,
although the climate is doubtfully favourable to the
growth of ordinary crops.

The striking features of the Laurentian plateau are
immeasurable lakes, large and small, with intervening
rounded rocky elevations, wooded, in their natural
conditions to the south, rising above the tree line to
the northward ; while in the far north, on both sides
of Hudson Bay, hills and valleys become eventually
characterized by grasses, mosses and lichens alone, constituting the great " barren lands " of North America.
The rivers and lakes are everywhere well stocked with
fish, while deer and moose in the southern parts, and to
the north the caribou, abound wherever the Indian
hunters have not followed them too closely.   Thus,
where the region can be entered without undue diffi-

culty, it has already become a much favoured resort of
the sportsman.

The name by which it is found convenient to refer
to this region, although derived from that of the
Laurentian system of geologists, must not be supposed
to imply that the rocks attributed to this system
occupy the whole area. There are, besides, several
wide and many narrower bands of Huronian rocks, as
well as outlying areas referred to the Lower Cambrian,
and more such exceptional areas of both kinds doubtless remain to be discovered. It must be understood,
too, that in employing the name Laurentian for the
most widely represented member of this great Archæan
plateau, this is done in a very general and inclusive
manner; the term "basement complex" is adopted
by some geologists in nearly the same sense. The
results of late investigations, particularly those of
Dr. F. D. Adams, seem to prove that in the Laurentian,
as thus understood, there is a distinctly stratified series
of limestones and other sedimentary rocks, all now
highly altered and crystalline, as well as another series
of more massive gneissic rocks, of which the apparent
bedding is really a foliation due to pressure, and which
frequently pass into granites by imperceptible grada-
tions. The first of these is known as the Grenville
series, and the second as the Fundamental Gneiss;
but even where most closely studied it has been found
impossible completely to separate the two series, except
quite locally, and thus, for the purposes of a general
geological map, both must of necessity be indicated by
a single colour. The complication alluded to is further
increased by the frequent occurrence of true eruptive
masses of granite or syenite of much later date, as
well as by that of numerous great areas of anorthosite,
a rock of the gabbro family, composed principally of
plagioclase felspar.

The Huronian rocks are as a rule darker in colour
and more basic in composition than those of the

Laurentian. They are often still distinctly bedded, although very frequently in the form of schists in which no true bedding is now distinguishable, and some of which have evidently resulted from the crushing of eruptive rocks. Over great areas the original material of the Huronian has clearly been in large part the result of volcanic eruptions of the time, but elsewhere, as on the north shore of Lake Huron, and in the Sudbury and Lake Temiscaming districts, it comprises thick masses of argillite, quartzite and quartz or jasper conglomerate. The rocks composing the Fundamental Gneiss often assume the physical relations of eruptives, of later date to the Huronian, along the lines of contact; but in how far this may really be the fact, and in how far this appearance may be attributed to a certain amount of re-fusion of previously existing basal rocks, has not been determined.

The Cambrian outliers date from a time subsequent to the main era of folding and crushing of the subjacent rocks, in comparison with which they are little disturbed. In the Labrador peninsula, they characterize considerable areas, and they form a border, tilted against the edge of the Laurentian plateau on the east side of Hudson Bay. These areas appear to represent the Animikie series of the Lake Superior section; well seen in similar relations in the vicinity of Port Arthur. These rocks are evidently separated by a vast period of unrepresented time from the Archaean below, and it is assumed that they may be attached with greatest probability to the base of the Cambrian, although no distinct fossils have been obtained from them. To the west of Hudson Bay and far to the north, are other outliers, largely composed of sandstones, conglomerates and traps, which resemble the Keweenawan or Nipigon of Lake Superior, somewhat later in age than the last. These also are provisionally classed as Lower Cambrian. The Nipigon rocks are well shown along the line of the Canadian Pacific railway to the east of Port Arthur.

Minerals of economic value in the Laurentian are almost entirely confined to the Grenville series, which yields, in various parts of its extent, mica, apatite (phosphate), graphite, iron and marble. The Huronian rocks are those in connection with which numerous important gold-bearing veins and deposits are now being found and opened up in western Ontario, as well as the nickel and copper deposits of Sudbury, iron and other metallic ores. In the Cambrian rocks near Port Arthur are silver-bearing veins, some of which have been extensively worked, and iron ores, while in the Labrador peninsula vast quantities of iron ores have been discovered in the same series.

Glaciation of Eastern Canada. Something may be added here on the events of the glacial period as affecting the eastern part of Canada as a whole, although many points connected with this particular period, still remain uncertain and the subject of debate. Like the Scandinavian peninsula, the Laurentian plateau at one stage in the glacial period apparently became the seat of a great confluent ice-sheet, which, when at its maximum, flowed down from it in all directions in general conformity with its main slopes. Climatic conditions and relatively local physical features may have conspired to render the discharge of glacier-ice more important in some directions than in others, and it is even possible that at no single time was the whole extent of the plateau equally ice-clad; but on the hypothesis above stated, it has been proposed by the writer to designate the ice-sheet of the Laurentian highlands, the Laurentide glacier. If several distinct centres of dispersion existed throughout, these may be named collectively the Laurentide glaciers. It is certain, at least, that such local centres must have grown up at the beginning of the period of cold, and that towards its close similar centres must have again remained as the last seats of ice action. Two of these have been defined by the existing marks

of rock striation and transport, one to the west of the
northern part of Hudson Bay, named by Mr. Tyrrell
the Keewatin glacier, the other, resulting from the
observations of Mr. Low, to the east of the bay, the
Labradorian glacier.

During the whole of this period the Laurentian
plateau was in the main an area of denudation. From
it the surface material was carried away in all directions, even to the northward, for there is absolutely no
evidence that any "polar ice-sheet" ever trenched
upon the continent of North America. The generally
bare ice-scored rocky surface of these highlands is
evidence of this denudation, while the existence of
broken, angular masses of unmoved local débris in the
central part of Labrador and in the central area of the
Keewatin glacier, shows that across these neutral
gathering-grounds, no ice ever passed.

As to the distance to which the solid glacier-ice
came southward from the Laurentian plateau, the
evidence is yet inconclusive, for it is at best a matter
of difficulty at the present day to definitely separate
true moraines from beaches upon which floating ice
impinged. Neither is it certain at how many times
or to what extent the glacial period was interrupted
by relatively warm epochs, but it may be stated that
the flora of at least one of these interglacial epochs, as
represented in the vicinity of Toronto, is such as to
indicate a climate fully as warm as that at present
existing, during which it seems improbable that much,
if any, glacier-ice could have persisted on the Laurentian highlands.

These problems cannot be discussed here, but it is
certain that at or about the decline of the glacial
period, the eastern part of Canada as a whole stood at
a relatively low level. Without quoting in detail the
heights to which the sea is known to have reached at
this time in various places, it may be stated that it
invaded the St. Lawrence valley as far at least as Lake

Ontario. Water-formed terraces are found at elevations of 230 feet in the Gaspé peninsula, to a maximum of 895 feet in the vicinity of Richmond, midway between Quebec and Montreal. Near Ottawa, the highest recognized shore-line is at 705 feet. The plains lying below these are floored by deposits holding marine shells of sub-arctic type, and these have been found at Montreal to a height of 560 feet. It is uncertain whether or not at the time of greatest subsidence the water covering the area of the Great Lakes stood at the same level as, and was in direct connection with the sea, but it is not improbable that what has been described as the "Iroquois beach," around these lakes, may actually form the continuation of a high marine shore-line. The question is complicated by the unequal amount of the subsidence and re-elevation to which the land has been subjected, a question requiring much further investigation.

As a result of the circumstances noted, the St. Lawrence plain, as far west as Ottawa and Kingston at least, is deeply covered by deposits due to the glacial period, including boulder-clay, Leda clay and an overlying Saxicava sand, the two last, often full of fossil shells of the period. In the coastal regions of the Maritime Provinces similar deposits occur, to which the same names have been extended, but to the west, around the Great Lakes, no marine forms have been found. The stony boulder-clay is there overlain by a fine clay named the Erie clay and by various newer sandy and clayey beds. All these deposits of the glacial period possess great importance in respect to the soil and agricultural character of the country over which they prevail, and in large part its exceptional fertility is to be attributed to them.

Before leaving this subject, it may be noted that the Laurentide glacier in the east, below the city of Quebec, never crossed the estuary of the St. Lawrence. A separate but smaller gathering-ground of ice existed

in New Brunswick and adjacent parts of the State of
Maine, which has been called the Appalachian glacier.
This has been very clearly shown by the work of Mr.
Chalmers. The Magdalen Islands, in the centre of the
Gulf of St. Lawrence, have never been glaciated, and
it is at least a matter of doubt whether any ice, except
that originating on the peninsula itself, ever passed
over Nova Scotia. Newfoundland similarly shows
evidence of having produced a local ice-sheet which
flowed away toward the sea in all directions along
the natural slopes. It is also found by Mr. Chalmers,
that when the Laurentide glacier invaded the lowlands
to the west of Quebec, the Appalachian glacier had
either greatly decreased or had vanished. There are
in this district two distinct boulder-clays, one referable
to each of these glaciers, and it is probable that interglacial beds will yet be found there.

The Interior Continental Plain. This is bounded on
the west by the Rocky Mountains, running about north-
northwest, and on the east by the edge of the Laurentian plateau, which, taking a more westerly direction
than the mountains, causes the gradual narrowing of
the intervening plain to the north. Thus on the 49th
parallel, here constituting the southern boundary of
Canada, the plain has a width of about 800 miles; but
it is reduced to less than 400 miles on the 56th parallel.
To the north of the 62nd parallel, it is very greatly
narrowed, and is broken by echelon-like flanking
ranges of the Rocky Mountains, but still further northward it again expands, and appears to have a width of
nearly 300 miles, where it terminates on the Arctic
Ocean.

The southern part of this great plain is not only the
most important from an economic point of view, but is
also that about which most is known. It includes the
wide prairie country of the Canadian west, with a spread
of about 193,000 square miles of open grass-land, an area
more than twice that of Great Britain. Beyond the

North Saskatchewan River, the plain becomes essentially a region of forest, with only occasional prairie tracts, such as those of the Peace River valley. By chance rather than by intention, the boundary line on the 49th parallel, to the west of the Red River, nearly coincides with the low watershed which separates the arid drainage-basin of the Missouri from that of the Saskatchewan and its tributaries, cutting off only 20,000 square miles from the Missouri slope. Another line, nearly coinciding with a second low transverse watershed, may be drawn on the 54th parallel. The watershed crosses this line several times, but in the main it may be taken as dividing the Saskatchewan system of rivers from those of the Mackenzie and the Churchill. The belt of country comprised between these latitude lines is 350 miles wide, with a total area of about 295,000 square miles.

The whole interior plain slopes eastward or northeastward, from the Rocky Mountains towards the foot of the Laurentian highlands, so that a line drawn from the base of the mountains near the 49th parallel to Lake Winnipeg, shows an average descent of over five feet to the mile, fully accounting for the generally rapid courses of the rivers of the region. There are, however, in the area to the south of the 54th parallel, two lines of escarpment or more abrupt slope, which serve to divide this part of the plain into three portions, and although such division is by no means definite, it may usefully be alluded to for purposes of description.

The first or lowest prairie-level is that of the Red River valley, of which the northern part is occupied by the Winnipeg group of lakes, its average elevation being about 800 feet above the sea, although gradually rising to the southward, along the axis of the valley, till it reaches a height of 960 feet about 200 miles to the south of the International boundary. Its area in Canada is about 55,000 square miles, including the lakes, and to the south of Lake Winnipeg it comprises

some 7,000 square miles of prairie land, which to the
eye is absolutely flat, although rising uniformly to the
east and west of the river.   This is the former bed of
the glacial "Lake Agassiz," the sediments of which
constitute the richest wheat lands of Manitoba.

The escarpment bounding this plain on the west, begins
at the south in what is known as "Pembina Mountain,"
and is continued northward in the Riding, Duck, Porcupine and Pasqua hills, which overlook Manitoba and
Winnipegosis Lakes, constituting the main eastern outcrop of the Cretaceous rocks of the plains.   From
this escarpment, the second prairie-level extends westward to a second and nearly parallel marked rise, which,
in general, is known as the Missouri Côteau.   The area
of this plain is about 105,000 square miles, of which
more than half is open prairie.   Its average elevation is about 1,600 feet, and its surface is more
diversified by undulations and low hills and ridges
than that of the last, while the river-valleys are often
deeply cut as well as wide.   The greater part of the
surface is well adapted for agriculture, although in
places the scarcity of trees constitutes a disadvantage.
The character of the soil is also more varied than that
of the lower plain.

The third and highest plain, lying between the last
and the base of the Rocky Mountains, may be stated
to have an average height of 3,000 feet, with an area,
between the parallels of latitude first referred to, of
about 134,000 square miles, of which by far the greater
part is almost absolutely devoid of forest, its wooded
area being confined to its northern and north-western
edges, near the North Saskatchewan River or its tributaries.   The surface of this plain is still more irregular
than that of the last, and it is evident that both before
and after the glacial period the denuding forces of rain
and rivers have acted upon it longer and more energetic-
ally.   Table-lands like those of the Cypress Hills and
Wood Mountain, must be regarded as outlying rem-

nants of an older plain of the Tertiary period, and the slopes and flanks of such outliers show that similar processes of waste are still in operation, adding to the length and depth of the ravines and "coulées," by which the soft Cretaceous and Tertiary rocks are trenched. The deposits of the glacial period, with which even this high plain is thickly covered, have tended to modify the minor asperities resulting from previous denudation. The soil is generally good, and often excellent, but large tracts to the south and west are sub-arid in character, and suited rather for pasturage than for agriculture.

Along the base of the Rocky Mountains is a belt of "foot-hills," forming a peculiar and picturesque region, of which the parallel ridges are due to the differing hardness of the Cretaceous rocks, here thrown into wave-like folds, as though crushed against the resistant mass of the older strata of the mountains.

Taken as a whole, the central plain of the continent in Canada may be regarded as a great shallow trough, of which, owing doubtless to post-Tertiary differential uplift, the western part of the floor is now higher in actual elevation than its eastern Laurentian rim. But although thus remarkably simple and definite in its grand plan, there are many irregularities in detail. The second prairie-level has, for instance, some elevations on its surface as high as the edge of the third plain, both to the west and east of the valley of the Assiniboine River, which, again, is abnormally depressed. It is not possible here to do more than characterize its features in a general way.

Ever since an early Palæozoic time, the area now occupied by the interior plain appears to have remained undisturbed, and to have been affected only by wide movements of subsidence or elevation, which, although doubtless unequal as between its different parts, have not materially affected the regularity of the strata laid down. Upon this portion of the continental platform, in

3

its eastern part, on Lake Winnipeg and its associated
lakes, Cambro-Silurian, Silurian and Devonian rocks
are found outcropping along the stable base of the
Laurentian plateau.   Following this line of outcrop
northward, the Devonian rocks gradually overlap those
of older date and rest directly upon the Archæan.
They continue to the Arctic Ocean and there occupy
a great part of the Northern Archipelago.   To the
south of Athabasca Lake they rest, without any
apparent angular unconformity, upon sandstones referred to the Lower Cambrian, to which allusion has
already been made, giving evidence, in the stratigraphical hiatus, of prolonged periods during Palæozoic time in which land as well as water existed in
some parts of the area.   On the western side of the
Great Plains the Palæozoic strata reappear crumpled
and broken in the Rocky Mountains, where the vast
crustal movements of the Cordilleran belt found their
inland limit.

These rocks consist, for the most part, of pale-grey
or buff, often magnesian limestones, along the eastern
outcrop, and from them Mr. Whiteaves has described an
extensive and somewhat peculiar fauna.   Some, at
least, of the Palæozoic formations represented, probably
extend beneath the entire area of the Great Plains,
but they are wholly concealed there by later strata of
Cretaceous age, consisting chiefly of clay-shales and
sandstones, generally but little indurated and flat-lying,
or nearly so.   The uniformity in the surface features
of this country is principally due to that of these
deposits, which, although since greatly denuded, have
worn down very equally, and have apparently never
been very long subjected to waste at a great height
above the base-level of erosion.   The whole area has
in fact been one rather of deposition than of
denudation up to a time geologically recent, and has
very lately been levelled up still further by the superficial deposits due to the glacial period.

The Cretaceous rocks are for the most part distinctly marine, although beginning with the Dakota sandstones, indicative at least of shallow seas, and from which, in the Western States, a considerable flora of land plants has been recovered. These are followed, in the eastern part of the plains, by the Benton shales, the Niobrara, largely calcareous and foraminiferal in some places, the Pierre shales, and lastly by the Fox Hill sandstones. Further west, in Alberta, a part of the Pierre (with probably also the upper portion of the Niobrara), is represented by the Belly River formation, with a brackish water fauna, and containing beds of coal or lignite. The Dunvegan series of the Peace River, to the north, similarly characterized, is perhaps somewhat older. The Cretaceous strata in fact change very materially in composition and character toward the Rocky Mountains, and when followed to the north, giving rise to the necessity for local names, and rendering a precise correlation difficult in the absence of connecting sections over great tracts of level country.

All the Cretaceous strata so far referred to belong to the later stages of that system, but in the foot-hills the earlier Cretaceous is represented by the Kootanie formation, holding coal, and reappearing as infolds in the eastern ranges of the Rocky Mountains. One of these is followed by the valley of the Bow River between Banff and Canmore, and affords both anthracite and bituminous coal.

Overlying the Cretaceous rocks proper, in considerable parts of their extent, particularly in Alberta, are those of the Laramie, which although perfectly conformable with the marine strata beneath, contain brackish water, and in their upper part entirely fresh water forms of molluscs, together with an extensive flora and numerous beds of lignite-coal or coal. As a whole, this formation may be regarded as a transition from the Cretaceous to the Tertiary, with a blending of organic forms elsewhere considered as characteristic of one or the other. The lower parts are undoubtedly most nearly related to the Cretaceous and particularly to the Belly River beds, which were laid down under similar physical conditions at an earlier stage. The remains of Dinosaurian reptiles are still abundant in these. The upper beds, constituting what was originally named the Fort Union group, with its local representatives under different names, is on the contrary more nearly allied to the Eocene.

A still later stage in the Tertiary is represented by beds of Oligocene, or early Miocene age, found particularly as an outlier capping the Cypress Hills. These have afforded numerous mammalian bones, described by Professor Cope and referred by him to the stage of the White River beds of the Western States.

The aggregate thickness of the Cretaceous strata of the plains, so far as known, may in the eastern part be stated as about 2,000 feet; in the west, in northern Alberta, it is about the same, but exceeds 2,500 feet in south-western Alberta, without including the Kootanie series of some 7,000 feet or more. The thickness of the Laramie is also great toward the Rocky Mountains, reaching probably 3,700 feet.

The Pliocene (with perhaps the latter part of the Miocene) appears to have been a time of erosion only in the area of the Canadian plains; wide, flat-bottomed valleys were cut out in the foot-hills, and to the east of these great tracts of country between the now outstanding plateaux must have been reduced to the extent of 1,000 feet or more in height.

Mineral fuels, in the form of coals and lignite-coals, constitute economically the most important products of the Cretaceous and Laramie rocks of the plains. To the south of the 56th parallel in Canada, an area of not less than 60,000 square miles is underlain by beds of such fuel. These consist entirely of lignite to the eastward, but on entering the foot-hills bituminous coals are found, and similar coals recur in infolded areas in the Rocky Mountains, together with anthracite.

Natural gas has been found in considerable quantities in borings in several places, but is not yet utilized. Great outcrops of Cretaceous sandstones saturated with tar or maltha occur along the Athabasca River, probably evidencing the existence of important petroleum deposits in the subjacent Devonian rocks. Salt springs appear on the borders of Manitoba Lake, and much further north in the Athabasca basin, but have been utilized to a very limited extent only. Gypsum also occurs in the Silurian and Devonian rocks along their eastern outcrop. Gold is washed from the sands of several of the larger rivers, but this, it is supposed, is for the most part derived by natural concentration from the drift deposits.

The Cordillera. Of this great mountainous region of the Pacific coast, a length of nearly 1,300 miles is included by the western part of Canada. Most of this is embraced in the Province of British Columbia, where it has a width of about 400 miles between the Great Plains and the Pacific Ocean. To the north, it is continued in the Yukon district of the North-West Territory, till it reaches in a less elevated and more widely spreading form, the shores of the Arctic Ocean on one side, and on the other passes across the 141st meridian of west longitude into Alaska. Its strongly marked features result from enormous crustal movements parallel to the edge of the Pacific, by which its strata have at several periods and along different lines, been crumpled, crushed and faulted. These movements having continued at intervals to times geologically recent, and the mountains produced by them still stand high and rugged, with streams flowing rapidly and with great erosive power down steep gradients to the sea.

Although preserving in the main, a general north-

north-westerly trend, the orographic features of this
region are very complicated in detail. No existing
map yet properly represents even the principal physical outlines, and the impression gained by the traveller
or explorer may well be one of confusion. Disregard-
ing, however, all minor irregularities, two dominant
mountain systems are discovered—the Rocky Mountains proper, on the east, and the Coast Range of
British Columbia, on the west.

The first of these, it has been proposed to name,
from an orographic point of view, the "Laramide
Range," as it is essentially due to earth movements,
occurring about the close of the Laramie period, and
rocks of that age are included in its flexures.
Although not quite continuous (for there are several
echelon-like breaks, in which one mountain-ridge
assumes the dominance previously possessed by
another), this range, beginning two or three degrees of
latitude to the south of the 49th parallel, forms the
eastern member of the Cordillera all the way to the
Arctic Ocean, which it reaches not far to the west of the
Mackenzie delta. It is chiefly composed of Palæozoic
rocks, largely limestones, and where it has been closely
studied is found to be affected by series of overthrust
faults, parallel to its direction, of which the easternmost
separates it from the area of the Cretaceous foot-hills.
Here the older rocks have been thrust eastward for
several miles over the much newer strata. The structure has as yet been worked out in detail only along
the line of the Bow River pass, by Mr. R. G. McCon-
nell. In width, this range seldom exceeds sixty miles.
The heights formerly attributed to some peaks appear
to have been exaggerated, but many points in its
southern part exceed 11,000 or 12,000 feet.

The Coast Range of British Columbia constitutes
the main western border of the Cordillera. Beginning
near the estuary of the Fraser River, it runs uninterruptedly northward, with an average width of about

100 miles, for at least 900 miles, when it passes inland beyond the head of Lynn Canal. This range is largely composed of granite, with infolded masses of altered Palæozoic strata. It is not, as a rule, so rugged in outline as the last, but its western side, rising from the sea, shows the full value of its elevation there, while its main summits often exceed 8,000 or 9,000 feet. Several rivers rising in the plateau country to the eastward, flow completely across this range to the Pacific, where the lower parts of their valleys, as well as those of many streams originating in the mountains themselves, in a submerged state constitute the remarkable system of fiords of British Columbia. Even in the arrangement of the islands adjacent to the coast, the further extension of these valleys and of others running with the range, may be traced, the evidence being of great subaërial erosion when the land previously stood at a higher stage. The cutting out of these deep valleys probably began in Eocene times, but was renewed and greatly increased in the later Pliocene.

Outside the Coast Range and in a partly submerged condition, lies another range, of which Vancouver Island and the Queen Charlotte Islands are projecting ridges. This stands on the edge of the Continental plateau with the great depths of the Pacific beyond it. The rocks resemble those of the Coast Range but include also masses of Triassic and Cretaceous strata which have participated in its folding, while horizontal Miocene and Pliocene beds skirt some parts of the shores.

In the inland portion of British Columbia, between the Coast and Rocky Mountain systems above particularly alluded to, are numerous less important mountain ranges, which, while preserving a general parallelism in trend, are much less continuous. Thus, in travelling westward by the line of the Canadian Pacific Railway, after descending from the Rocky Mountain summit and crossing the Upper Columbia valley, the

Selkirk Range has to be surmounted. Beyond this, the Columbia on its southward return is again crossed, and the Gold Range is traversed by the Eagle Pass before entering the Interior Plateau of British Columbia, which occupies the space remaining between this and the Coast Range. The system of ranges lying immediately to the west of the Rocky Mountains proper, notwithstanding its breaks and irregularities, is capable of approximate definition and its components have been designated collectively the Gold Ranges. Further north, it is represented by the Cariboo Mountains, in the mining district of the same name. The highest known summit of this system is Mount Sir Donald, 10,645 feet, one of the Selkirk Mountains.

This mountain system is believed to be the oldest in British Columbia. It comprises Archæan rocks with granites and a great thickness of older Palæozoic beds, much disturbed and altered.

The Interior Plateau constitutes an important physical feature. Near the International boundary it is terminated southward by a coalescence of rather irregular mountains, and again, to the northward, it ends about latitude 55° 30′ in another plexus of mountains without wide intervals. Its breadth between the margins of the Gold Ranges and the Coast Range is about 100 miles, and its length is about 500 miles. It is convenient to speak of the country thus defined as a plateau, because of its difference, in the large, from the more lofty bordering mountains. It comprises the area of an early Tertiary denudation-plain (or peneplain) which has subsequently been greatly modified by volcanic accumulations of the Miocene, and by river-erosion while it stood at a considerable altitude, in the Pliocene; but its true character as a table-land is not obvious until some height has been gained above the lower valleys, where the eye can range along its level horizon-lines. It is highest to the southward, but most of the great valleys traversing it are less in elevation than 3,000 feet above the sea. To the north, and particularly in the vicinity of the group of large lakes occurring there, its main area is less elevated than 3,000 feet, making its average height about 3,500 feet.

Beyond this plateau to the north, the whole width of the Cordillera, very imperfectly explored as yet, appears to be mountainous as far as the 59th parallel of latitude, when the ranges diverge or decline, and in the upper basin of the Yukon, rolling or nearly flat land, at moderate elevations, again begins to occupy wide intervening tracts.

As a whole, the area of the Cordillera in Canada may be described as forest-clad, but the growth of trees is more luxuriant on the western slopes of each of the dominant mountain ranges, in correspondence with the greater precipitation occurring on these slopes. This is particularly the case in the coast region and on the seaward side of the Coast Range, where magnificent and dense forests of coniferous trees occupy almost the whole available surface. The Interior Plateau, however, constitutes the southern part of a notably dry belt, and includes wide stretches of open grass-covered hills and valleys, forming excellent cattle ranges. Further north, along the same belt, similar open country appears intermittently, but the forest invades the greater part of the region. It is only toward the Arctic coast, in relatively very high latitudes, that the barren Arctic tundra country begins, which, sweeping in wider development to the westward, occupies most of the interior of Alaska.

With certain exceptions, the farming land of British Columbia is confined to the valleys and tracts below 3,000 feet, by reason of the summer frosts occurring at greater heights. There is, however, a considerable area of such land in the aggregate, with a soil generally of great fertility. In the southern valleys of the interior, irrigation is necessary for the growth of crops.

The geological structure of the Cordillera is extremely complicated, and it has as yet been studied in detail over limited tracts only. There have been no appropriate terms of comparison for the formations met with, and these it has consequently been necessary to investigate independently by the light of first principle. The difficulty is increased by the abundance of rocks of volcanic origin referable to several distinct periods, resembling those of the Appalachian mountain region, though on a vastly greater scale, and like them, almost entirely devoid of organic remains. The recognition, early in their investigation, of the originally volcanic nature of a large part of the rocks, has rendered it possible, however, to understand the main geological features, which at first appeared to present an almost insoluble problem.

Rocks referable to the Archæan obtain a considerable development in British Columbia, but it has not so far been found possible to recognize definitely the Laurentian and Huronian systems. Where they have been noted and examined, chiefly in the Gold Ranges and Interior Plateau, they have been distinguished as the Shuswap series. They include rocks resembling the Fundamental Gneiss of the east in character and composition, together with crystalline limestones, quartzites and gneissic rocks like those of the Grenville series and evidently representing metamorphosed sediments. At this distance from the typical developments of the Laurentian and Huronian, it is not to be expected that any precise parallelism in mineral composition and degree of alteration can be established, but that these rocks really are Archæan, has been determined by their unconformable infraposition to the lowest Palæozoic strata.

Above the Shuswap series, in the Rocky Mountains, Gold Ranges and elsewhere, is a great thickness of Cambrian strata, or of Palæozoic rocks at present collectively classed under this system. Fossils referable to the *Olenellus* zone, have been found about half-way down in these strata in the Rocky Mountains, but from the lower conformable mass of beds, no recognizable organic forms have yet been obtained. A great thickness of contemporaneous volcanic material is generally included in the Cambrian. The Cambro-Silurian and Silurian, on the evidence in each case of a few characteristic fossils, are known to exist in the western part of the Rocky Mountains proper, and far to the north, on the Dease River (near lat. 60°), an interesting Graptolitic fauna of Trenton age has been found. The Devonian has not been distinctly recognized.

In the Rocky Mountains, the Carboniferous is largely represented, chiefly by massive limestones, and the fossils found in these pass down to a stage which has been characterized as Devono-Carboniferous. No single trace of the flora of the Carboniferous period has yet been discovered in the western regions of Canada. In the Interior Plateau and along the coast, the Carboniferous consists below of volcanic accumulations and quartzites and above of limestones, some of which are largely foraminiferal and composed of *Fusulina* and *Loftusia*.

The Triassic, in the southern part of the Rocky Mountains proper, is represented by red sandstones, the deposits of an interior Mediterranean of the period. To the west and north, it becomes a marine formation, with peculiar fossils of the " Alpine Trias " type, but over large areas it consists almost entirely of contemporaneous volcanic accumulations.

In a few places in the southern part of British Columbia, this formation appears (following the views of Prof. A. Hyatt on its fossils) to pass up into the Jurassic; but the next important series of beds, succeeding a very great stratigraphical break, is the Cretaceous. Rocks of the Earlier Cretaceous (Kootanie and Queen Charlotte Island formations) occur in places in the Rocky Mountains and throughout

British Columbia as far as the coast, also northward to the Porcupine River, between latitudes 67° and 68°, in the Yukon District. Newer Cretaceous rocks are developed particularly in Vancouver Island, where they constitute the productive coal measures. In the Crow's Nest Pass region and elsewhere in the Rocky Mountains, as well as in the Queen Charlotte Islands, the Earlier Cretaceous rocks contain abundance of good coal. All the strata of the Cretaceous period are more or less tilted and folded, and are evidently prior in date to the last great orogenic movements of the Cordillera. Evidences of contemporaneous volcanic action are again abundant in some parts of the extent of the Cretaceous.

Rocks referable to the Laramie or transition period between the Cretaceous and Tertiary, are found in the Yukon district and in the vicinity of the Fraser delta, holding lignite-coals and numerous remains of plants. Beds assigned to the Oligocene and Miocene are also well developed in the southern part of the Interior Plateau of British Columbia, where the latter period has been an epoch of notable volcanic eruptions, producing both effusive and fragmental rocks, but toward the close flooding large tracts with basaltic flows. Traces of similar volcanic activity, of the same date, are found in the Queen Charlotte Islands and in Vancouver Island. The Pliocene was chiefly a time of erosion, but deposits referred to this period are not entirely wanting.

Until the completion of the Canadian Pacific transcontinental railway, the west coast of Canada was a remote region, accessible with difficulty; but long before this, coal had been successfully mined in Vancouver Island, and in 1858 and succeeding years, the discovery and working of placer gold deposits brought the then isolated colony of British Columbia into considerable prominence. From the time of the quarrel with Spain on the Nootka question, in 1790, little had been heard of the region, which remained unprized and suffered naturally in consequence when the " Oregon " boundary was settled with the United States.

The yield of alluvial gold reached its maximum in 1863, and thereafter, as in all mining regions of this kind, began steadily to decline. But of late years, and consequent on the introduction of means of communication, a remarkable development has begun. The conditions in this disturbed region of the continent are so varied, that it is difficult to name many metallic minerals which do not occur in it; but silver, gold, lead and copper have already begun to be produced in important quantity where certain districts in its southern part have been prospected, while the placer mining of gold had extended very far to the north in the Yukon valley and to the border of Alaska. Coal mining has continued uninterruptedly since its first beginning, and the product is now exported to many countries on the Pacific seaboard, but more particularly to California.

The Cordilleran region of Canada is undoubtedly destined to become one of the most important mining countries of the world, and probably before many years the value of its output will more than equal that of all other parts of the Dominion combined.

Glaciation of Western Canada. Like the eastern part of Canada, the western has been largely affected by the events of the glacial period. Most of the superficial deposits can be explained only by reference to this period, and to it also the diversion of many rivers and streams and other important changes are due. It is not yet possible to give a connected account of these events which will meet with general agreement, but, as in the east, the main facts have already been made sufficiently plain.

At an early time in the glacial period, the Cordillera, standing probably at a relatively high elevation, became covered by a confluent ice-sheet, extending approximately from latitude 48° to latitude 63°, with a total length, at its maximum, of some 1,200 miles. The form of the surface prevented the ice from discharging in all directions like that of Greenland, and forced the bulk of the outflow to move south-eastward and north-westward, in conformity with the direction of the ruling mountain ranges, from a central neutral gathering-ground or névé, situated approximately between the 55th and 59th parallels. The southward-moving portion of the great glacier filled the Interior Plateau of British Columbia, while its opposite extremity in the main flowed into the Yukon basin. Smaller streams from the main mass undoubtedly crossed the Coast Range by transverse valleys, to reinforce secondary, but large glaciers, which reached the sea to the south and north of Vancouver Island, while others extended through the Bow River valley and similar depressions in the Rocky Mountains to the western margin of the Great Plains.

This Cordilleran glacier, as shown by late observations in the western part of the Great Plains, was the first to affect that region, and may perhaps prove to be the first notable ice-cap developed during the glacial period in North America. At a later time it became gradually very much reduced, but subsequently, at least once, again extended to dimensions in some places approaching those first held by it. Rock striation and the transport of erratics, show that the southern part of the Cordilleran glacier, when at its maximum, passed uninterruptedly over projecting points between 6,000 and 7,000 feet in height above the sea.

In the opinion of the writer, there is evidence such as to render it probable that the first retreat of the Cordilleran glacier was contemporaneous with a depression of the Cordillera to the amount of 4,000 feet or more, enabling water at the level of the sea to reach such elevations on both sides of the Rocky Mountains.

Subsequently, during the second spread of the glacier-ice, the land is supposed to have risen; and at a still later date, there is fairly conclusive evidence to the effect that it stood about 2,500 feet lower relatively to the sea than it now does.

The above hypotheses are stated under all reserve and subject to further enquiry, but the main facts and the evidences of glaciation of a very pronounced type, rock-scoring, boulder-clay, moraines, terraces, kames and eskers are abundantly evident throughout the greater part of British Columbia and the Yukon district.

In the area of the Great Plains, as above noted, the first recognized evidences of glacial conditions are those connected with the eastward spread of comparatively limited tongues of glacier-ice from the Rocky Mountains, and the deposit, on the western plains, of boulder-clay and rolled gravels attributed to what it has been proposed to name the Albertan stage. Subsequently, at least two more distinct boulder-clays, separated by important interglacial deposits, have been laid down over the whole western part of the Great Plains, ending above in silty, sandy and gravelly beds, with large scattered superficial erratics. In connection, doubtless, with one of these boulder-clays, is the remarkable monument of the Glacial period known as the Missouri Côteau (crossed by the Canadian Pacific railway west of Parkbeg station), which may be regarded either as part of a continental moraine or as the marginal accumulation of an ice-laden sea. These later boulder-clays differ from those of the Albertan stage in being largely composed of debris of the Laurentian and Huronian rocks and Palæozoic limestones found in places on the eastern side of the interior continental plain. The direction of transport of these erratics has been from the north-east or north-northeast. The manner of their carriage and that of the deposition of the beds in which they occur, is still a subject of discussion; but there can be no doubt that great relative and absolute changes in elevation have occurred in the region, while terraces occurring at heights of 5,300 feet in the Porcupine Hills, near the base of the Rocky Mountains, show that that part of the plains at least was flooded to a corresponding height, which is practically identical with that previously found in the mountain region of British Columbia. To the writer there appears to be strong presumptive evidence that this water was in more or less direct communication with that of the sea, but other hypotheses have been advanced, and it remains for future investigation to determine that which may eventually prevail. It is with hesitation that we are prepared to admit great succeeding changes in level of large areas of the continent, but the alternative explanations, attributing as they must the most extraordinary effects to glacier-ice, seem to present at least equal difficulty.

CHAPTER I.—APPENDIX A.

Survey of Tides and Currents in Canadian Waters.

By W. Bell Dawson, Ma.E., Assoc. M. Inst. C.E., F.R.S.C.

———

THE importance of establishing stations for tidal observations on which tide tables for Canadian waters could be based, was discussed in 1884, in a paper read by Dr. A. Johnson, of McGill College, when the British Association held its meeting in Montreal. A Committee of the Association was then appointed to collect information, and to make representations to the Government regarding it. Of this Committee, Dr. Johnson was chairman. The Montreal Board of Trade were at the same time considering the question of wrecks caused by unknown currents; and they concurred in addressing a strong memorial on the subject to the Dominion Government. Ship owners and masters of vessels were also practically unanimous as to the pressing need for knowledge on the subject of tides and currents.

After numerous representations and negotiations, a beginning was made in 1890, when $2,000 was voted for the purpose. Preliminary investigations had been made by the late Lieutenant Gordon; and these were followed up by the late Mr. C. Carpmael, director of the Meteorological Service. Under Mr. Carpmael's supervision several tide gauges were placed before the close of 1893, and equipped with self-recording instruments. These gauges may be considered as tentative. Their localities were well chosen; but their scales had to be altered to correspond with the range of the tide,

or their construction improved, before records of practical value were obtained from them.

In December, 1893, the present writer was appointed
to the charge of the work; and it was also decided to
commence in the following season an investigation of
the currents. The survey as thus constituted, is now
included in the Technical Branch of the Department of
Marine and Fisheries, with Mr. W. P. Anderson, Chief
Engineer, at its head. The survey consists of two
divisions; namely, (1) Tidal observations, and the
computation and publication of tide tables; and (2)
A survey of the currents on our eastern coasts. For
three years these two branches of the work have been
carried on simultaneously, and important results have
already been obtained.

The Tides. The tidal division of the work presents
exceptional difficulties, on account of the wide varia-
tion in the range of the tide on our coasts, and also
because the tide gauges have to be specially designed
with a view to heating in winter, to prevent them
from freezing up. In the Bay of Fundy, the average
range is over thirty feet; while the Atlantic tide at
Halifax has a range of six feet. This Atlantic tide,
after entering the Gulf of St. Lawrence is almost
effaced on its way across, but regains its height at the
entrance to the St. Lawrence River, and proceeds up
the estuary with ever increasing height, till at Quebec
it attains a range of nearly twenty feet. In the Gulf
of St. Lawrence there are also serious complications
which occasion in some regions a marked diurnal
inequality; but this is inappreciable in the estuary
of the St. Lawrence, where the regularity in the progress of the tide is in marked contrast with the
irregularities in the open Gulf.

Principal Tidal Stations. In these circumstances
it was necessary to have a number of tidal stations at
which continuous observations could be obtained
throughout the year. Positions for these were selected

with much consideration ; so that they might serve as
principal stations from which tidal differences for
intermediate places could be determined. Several of
them had also to be erected on rocky coasts where no
wharves or other facilities existed. The positions
chosen were as follows : *St. John, N. B.*, in the Bay of
Fundy ; *Halifax, N. S.*, to secure the Atlantic tide ;
St. Paul Island, in Cabot Strait, between Cape Breton
and Newfoundland, and *Forteau Bay*, in Belle Isle
Strait, to command the two entrances to the Gulf of
St. Lawrence ; *South-west Point of Anticosti*, at the
entrance to the St. Lawrence River ; *Father Point*, the
Pilot station ; and *Quebec*. Tide gauges were erected
at these positions in 1893 to 1895 ; equipped with self-
recording instruments, sight-gauges for the determina-
tion of datum, and barographs where required. They
are all provided also with heating in winter.

The tidal recording instrument has the disadvantage
inherent in all types of tide gauges, that the driving
clock is an integral part of the instrument, and when
it needs cleaning or repair, the whole instrument has
to be removed. The most serious interruptions have
been due to this cause, especially at isolated stations
with which there is no communication during five
months of the year. A new type of recording instrument has therefore been devised by the writer, in
which the driving clock is removable, and can be
replaced by a duplicate in a few minutes. It is also
furnished with interchangeable gearing, giving four
different scales corresponding with ranges of 9, 18, 27,
and 36 feet. Where the tide has so great a variety in
its range, this will obviate the necessity of returning
the instrument to the makers in Britain for change of
scale, as has been necessary in the past.

Tide Tables. Formerly such tide tables as were
published for Canada, were based upon constant differences with distant ports, usually on the other side of
the Atlantic. The tide tables issued by this survey

give for the first time reliable tables based upon direct observation, which includes the height as well as the time of the tide. As early as 1891, tide tables were issued for Halifax which were based upon old records of 1860-61. It was found difficult to obtain satisfactory circulation for these, and latterly the tables have been supplied to the leading British and Canadian Almanacs for publication. The available records have already been utilized as a basis for tide tables as follows :

FOR 1896. *Halifax.* Old records of 1860-61.

Quebec. Record of November, 1893, to January, 1895. With tidal differences for the Atlantic coast of Nova Scotia, and a portion of the Lower St. Lawrence.

FOR 1897. *Halifax.* Old records of 1860-61 and 1851-52, with tidal differences for Nova Scotia.

Quebec. Record of November, 1893, to January, 1895. Tidal differences extended to include the whole tidal portion of the St. Lawrence, from Three Rivers to Anticosti, a distance of 420 miles.

Ste. Croix Bar, above Quebec ; where there is least water at present, until the dredging of the channel is completed.

Father Point. Computed by difference from Quebec, based on simultaneous record from the gauges.

Charlottetown. Record, June to November, 1896.

Pictou, N. S. Record, June to November, 1896.

FOR 1898. Tide tables as above; the basis of the Quebec tables being extended from November, 1893, to January, 1896. An extensive series of tidal differences for ports in the Gulf of St. Lawrence are in preparation, based upon simultaneous observations during several months in the summer season of 1896, taken at a number of carefully selected points.

Survey of the Currents. This was undertaken in the direct interest of shipping ; and as two leading steamship routes traverse the Gulf of St. Lawrence, that region was first examined. Although in the es-

tuary of the St. Lawrence, stronger tidal streams occur,
they are more nearly along shore, and less liable to set
a vessel out of position. It has not been possible to
provide a special steamer for the purposes of this investigation; but one of the coaling and buoy steamers,
of the lighthouse supply service, has been placed at the
disposal of the survey, for three months in the summer
seasons of 1894, 1895, and 1896. The general method
adopted, was to anchor the vessel at carefully selected
stations, and to determine the strength and direction of
the current from it, as a fixed point. The chief difficulties lay in holding so unwieldy a vessel at anchor,
and in want of means to provide suitable anchorage
appliances. Anchorages had to be made under all conditions, ranging from 30 fathoms depth on flat rock
bottom, to 250 fathoms in soft mud. The direction
and velocity of the currents were obtained by an attached float, and by a current meter registering elec-
trically on board. As electric meters had been little
used at sea, much trouble arose from short-circuiting,
owing to the higher conductivity of sea water; but
this was overcome. By their use a continuous record
day and night, could be obtained when desired. It
was also important to obtain a knowledge of the un-
der-currents, as they indicate the character of the general circulation much more correctly than the surface
current, which is much affected by wind drift. The
under-current observations usually extended from 10
to 40 fathoms; as below 60 fathoms there was
seldom any appreciable current. Continuous meteorological observations were also taken on board for
comparison; and the tidal data required were furnished
by the tide-gauges. The density of the water furnished a valuable indication wherever the influence of
the St. Lawrence water could be detected. For this,
hydrometers of special range were used. The temperatures gave less definite indication of the direction of
the movement of the water. It was found that the

coldest water (30° to 34° F.) occurred at a depth of
about 50 fathoms, and below this to 200 fathoms in
the deeper parts, the temperature was again higher.
It was, therefore, necessary to employ deep-sea invert-
ing thermometers, as well as registering thermometers
nearer the surface.

In the first season, the currents in the two entrances
to the Gulf were examined; namely, Cabot Strait and
Belle Isle Strait, by which the Gulf communicates with
the ocean.   The current in Belle Isle Strait was found
to be tidal, with a flow nearly equal in each direction.
The accepted theory of a constant inward current thus
proved to be unfounded and misleading. In the second
season, the constant outward current at the mouth of
the St. Lawrence between Gaspé and Anticosti was examined; and it was found that the St. Lawrence water
takes a south-easterly direction across the Gulf towards
Cape Breton, and passes out of the Gulf at Cabot
Strait, as evidenced by the lower density of the water.
In the third season, an examination was made of the
north-eastern half of the Gulf, from Anticosti to Belle
Isle Strait; and some conclusions as to the general character of the Gulf currents and the circulation of the
water were arrived at.

The methods used and the results obtained are fully
described in the Reports of Progress which have been
issued.   These have been distributed as widely as
possible; in particular they have been supplied to the
captains of ten steamship companies, mostly Trans-
atlantic lines, whose steamers traverse the Gulf.   The
currents in the Bay of Fundy have not yet been examined.   Investigation is also necessary on the south
coast of Newfoundland, where several wrecks have
been caused by the indraught into the larger bays
on the coast.   After the examination of the regions
traversed by the leading steamship routes, which is the
most urgent, there will still remain much to be done
before the information is complete.

By means of the tidal observations, the elevations of mean sea level and of low water are being correctly determined in our principal ports, which will be of much value in furnishing datum planes of reference. The datum planes in our cities are still in a position which is far from satisfactory.

In the Bay of Fundy and on the St. Lawrence, there are exceptionally good opportunities for the study of the effect of wind and atmospheric pressure on the tides; a subject to which a Committee of the British Association is now giving special attention. The simultaneous observations obtained last season on self-recording gauges at ten different points on the Atlantic coast and in the interior of the Gulf, furnish excellent material from which to investigate the secondary tidal undulations of about twenty minutes period, which at times are very marked. The requisite data for such collateral investigations are being collected; but as both divisions of the survey are carried on by the Engineer-in-charge with the help of two assistants, and on an annual expenditure limited to about $12,000, it has not been possible to overtake more than the immediate practical requirements of the work.

CHAPTER I.—APPENDIX B.

Photographic Surveys.

By E. Deville, Surveyor-General of Canada, Ottawa.

THE application of photography to surveying has received a wide extension in Canada, far exceeding anything attempted in any other country. The results may, on that account, prove of interest.

It was first employed here in 1887, when the surveys of Dominion lands were extended to the Rocky Mountains. In the prairies, the operations of the land surveys are limited to defining the boundaries of townships and sections: these lines form a network over the land by means of which the topographical features, always scarce in the prairies, are sufficiently well located for general purposes. In passing to the mountains, the conditions are entirely different. The topographical features are well marked and numerous; the survey of the section lines is always difficult, often impossible, and in most cases useless. The proper administration of the country required a tolerably accurate map, and means had to be found to execute it rapidly and at a moderate cost. The ordinary methods of topographical surveying were too slow and expensive for the purpose; rapid surveys, based on triangulation and sketches, were tried and proved ineffectual; then photography was resorted to.

Landscape sketches have long been used by topographers for recording observations and as a help for plotting the plans, but it was not until 1849 that

Laussedat, an officer of engineers, tried to plot plans from accurate perspectives drawn with a camera lucida. His paper on the subject, written in 1850, was published in 1854. Shortly after, he substituted photography for the camera lucida. He gave full particulars of the method in various papers, and his work was so complete that little has been added to it since. Many modifications have been suggested, but the soundness of Laussedat's views is shown by the fact that wherever photographic surveys are now being made, they are executed by the application of the principles which he has laid down.

A survey of any kind is made by measuring angles and distances. An accurate plan can be prepared from angular measurements alone, but it is not complete until the scale has been fixed by measuring one or more distances. The rôle of photography is to dispense with angular measurements on the ground: they are deduced from the photographs. A large portion of the work is thus transferred to the office instead of having to be executed in the field, where the expenses of a survey party are considerable.

The photographic image produced by a good rectilinear lens is a true perspective, the "principal point" being the foot of the perpendicular drawn from the second nodal point to the face of the photographic plate. This perpendicular is the "distance line" and should be equal to the equivalent focal length of the lens when properly focussed upon distant objects. The vertical and horizontal lines drawn through the principal point are the principal and horizon lines of the perspective. The azimuth of the principal point being known, the azimuth of any other point of the photograph as well as its angular distance from the horizon are obtained by simple geometrical constructions and transferred to the plan.

On the Canadian surveys, a primary triangulation is first executed by the usual processes: then comes the photographic surveyor. His equipment consists of a three inch transit theodolite and two cameras, one for himself and one for his assistant. With the theodolite, he locates his camera stations, makes secondary triangles where necessary and measures the azimuth of at least one well defined point for each photograph. From this azimuth, the direction of the principal point is deduced.

The camera is a rectangular metal box $4\frac{3}{4}$ x $6\frac{1}{2}$ inches provided with levels: the lens is a Zeiss anastigmat. By means of a screw, the photographic plate is brought in contact with the edges of the box so that the focal length is invariable. The horizon and principal lines are indicated by notches in the edges. The metal box is itself enclosed in a mahogany box to which the triangular base with levelling screws is attached. The whole instrument is extremely strong and compact, and can stand much rough usage without being put out of adjustment. Each single plate holder bears a number which is inscribed in pencil on the plate when it is inserted in the holder. The camera and twelve holders pack in a leather case with shoulder straps: the weight is fourteen and a half pounds.

By using isochromatic plates and a deep orange screen, the bromide film is acted upon almost exclusively by the yellow rays and the image of distant points is obtained with great distinctness, the effect of the blue haze or "aerial perspective" as it is called, being entirely absent.

The half plate negatives are enlarged on extra thick bromide paper to a little over twice their size; with care in drying, there is no distortion.

Having plotted his triangulation and the camera stations, the surveyor takes a photograph and finds the azimuth of the principal point, which he draws upon the plan. At a distance equal to the enlarged focal length, he draws a perpendicular which is the ground line. He repeats the construction for another photograph covering the same ground. Then he places the two photographs before him and selects a number of points which can be identified upon both views: to each point he assigns a number. The directions are next plotted on the plan, the altitudes being obtained by geometrical constructions. The process is much the same as with the plane table.

Up to 1892, the photographic surveys were confined to the Rocky Mountains, in the vicinity of the Canadian Pacific Railway; at the end of that year, they covered about 2,000 square miles. In the same year, an International Boundary Commission was appointed to examine the country along the boundary between Canada and the United States Territory of Alaska. The Canadian Commissioner, Mr. W. F. King, decided to carry out his share of the work by photography. In 1893 and 1894, his parties surveyed about 14,000 square miles.

Irrigation surveys were commenced in 1894 in the south-westerly part of the North-West Territories, where the rainfall is not quite sufficient for agricultural purposes. In addition to the gauging of streams, the establishment of bench marks, etc., it is necessary to ascertain the catchment areas and to define the sites best adapted for reservoirs. For this purpose photography has again been resorted to in the foot hills and on the eastern slope of the mountains. It has, in this case, a peculiar advantage. Whether or not a site is a favourable one for a reservoir cannot be known until the plan has been partly plotted. It must be possible to bring water to the proposed place and to run it off; the capacity must also be adequate. If favourable, a detailed survey of the site is required. With the ordinary surveying instruments, a preliminary survey has to be made; if, after plotting it, the site is found favourable, the topographer has to go over the ground a second time to make a detailed survey; or, the

whole of the work may be executed at once, with the
contingency that the detailed survey may turn out
useless. With the camera, the plan may be plotted so
far, and so far only, as required; the photographs
which furnish a general plan, can be made to give all
the detail wanted without going again into the field.
Whether the site is a good one or not, there is no
labour wasted.

CHAPTER I.—APPENDIX C.

The Canadian Hydrographic Survey.

By Wm. J. Stewart, Officer in Charge of Canadian
Hydrographic Survey.

———

THE Hydrographic Survey of Canada was inaugurated under the late Honourable A. W. McLellan,
by Staff Captain J. G. Boulton, R.N., in 1883. It was
undertaken because of the many unheard of dangers,
constantly being reported, that were causing heavy
losses both in lives and shipping upon the Great
Lakes.

Previously mariners had only the small scale charts
made by the late Admiral Bayfield in 1817-1822.
Considering the time taken over them and the facilities he and his large staff of assistants had, the charts
are really excellent; but no attempt was made to
delineate the shoals and dangers. Very few soundings
were taken, and therefore the charts lack their chief
usefulness.

In Canada we have no Geodetic Survey to form a
base for the Hydrographic work (if we except some
little assistance obtained from the United States Coast
and Geodetic Survey), and therefore our work has to
be more carefully performed without attempting the
accuracy of a Geodetic Survey. Our instruments are,
comparatively speaking, small. For triangulation we
use an eight-inch transit theodolite, with smaller
instruments for details. Then we have sextants of
various kinds, from the finished observing sextant to
the box and sounding sextants, box chronometers and
chronometer watches.

A few words as to the observations most commonly taken. There is that for azimuth, either on the sun or Polaris. That on the sun is taken with the transit theodolite, to give both the altitude and the horizontal reading of the sun, both early in the morning and late in the afternoon when the exact time is not very important. Sometimes, however, the sextant and artificial horizon are used for the altitude to be taken, at the exact instant another observer takes a horizontal reading with a theodolite. The azimuth by Polaris is a well-known observation.

Latitude is obtained either by using the sextants or transit theodolite : the observations are essentially the same, and consist in obtaining the altitudes of certain stars crossing the meridian at a known time. We usually combine results from circum meridian altitudes of several south culminating stars with results from many observations of Polaris. The means of several nights' work give very good results.

Observations for errors of the chronometers are taken with a sextant (supported on a stand), equal altitudes before and after noon giving the chronometer error at noon. Usually a large number of observations are taken in the morning, and as many equal altitudes as possible taken in the afternoon. Equal altitudes of stars may be used to give the error at midnight.

For a season's work the shore between some two salient points is usually selected. This being done, the main stations are erected as most convenient, but they can seldom be placed inland. If the shore and off-lying islands be suitable the triangulation is a simple matter, but if the shore be nearly straight and have no islands off it, recourse must be had to buoys moored short, as has been done along a good portion of the north shore of Lake Erie. By carefully placing the buoys, paying great attention to the mooring, and selecting good quiet days for observing the angles

from the main stations to these buoys, very little error
is introduced. This error is ascertained by measuring
a check base in some convenient locality. Bases with
us are not very long, from 2,000 to 4,000 feet, and are
measured upon level ground with a carefully compared
chain. The triangles are enlarged from this in the
usual way.

Between the main stations smaller ones are erected
as found needful, say about every half mile, to fix
points on the shore, and to aid in the location of the
soundings. Between the stations the shore line is
traversed, fixed, and all prominent buildings, trees and
hills cut in. Next in order, after these stations are
plotted, comes the sounding, which is usually done in
three stages—boat sounding, ship sounding and careful
examination of suspicious casts.

Boat sounding is most conveniently carried on from
whale boats, with four or six men to row, a man to
sound from the bow, and an officer, who has with him
a sounding sextant, station pointer, and a plan showing all the stations in his locality properly plotted, and
lines drawn parallel to one another at convenient
distances apart and at right angles to the general trend
of the shore. The soundings are taken as fast as the
man can sound and only shoal or representative soundings recorded by the officer. The position of the boat
is fixed from time to time by stopping and measuring
two angles to three stations ashore. These angles are
laid off on a station pointer and the position plotted on
the plan, in the boat, so as to keep as near the ruled
lines as possible. This boat sounding is carried out
from shore to a safe depth so that the steamer may
move in safety.

Ship sounding is the work in deeper water and
further removed from shore. The soundings are taken
from the deck (there being various appliances to
obtain these without stopping the steamer each time,
as would be necessary if we only had a measured line).

The sextant angles (to fix the sounding) are taken on deck by two officers at the instant the lead drops. The fixing is at once plotted on the sheet and the vessel kept as nearly as possible upon parallel lines, which run out as far as it is possible to fix.

Fixing soundings out of sight of land requires much care and can hardly be explained here. No such work has yet been attempted by us on the Great Lakes; occasional lines carried by us across the lakes reveal no shoals to require any very extended examination of the lakes out of sight of the land.

The examination of suspicious casts is the important work to which all the rest is preliminary. The lines give cross sections of the bottom, and we expect the lines to cross some part of the shoals of which we thus get a warning, by having a cast more or less shoaler than the one on each side of it. These casts are all inked in upon the boat plans, a boat visits the locality, drops over small buoys, as the officer finds necessary, crosses and recrosses the suspected ground till he is satisfied he has found the *least* water on the shoal, which he fixes. This shoal hunting is a very tedious task, and requires no end of patience and a considerable amount of skill both in the officer and sounder. The shoals located, it is the business of the survey to give when possible "leading marks" to assist sailors in avoiding the dangers.

The nature of the bottom is obtained by examination of a small quantity brought up by a priming of tallow upon the end of the lead. At times, with a different apparatus, larger quantities of the bed are obtained, but nothing of any great interest has been obtained. The marine interests are usually served by knowing whether the bottom is hard or soft; the fishing interests require a little more knowledge, such as gravel, solid rock, boulder, sand, clay or mud bottoms. The chart is now made up, the shore line is in and the soundings taken; next comes the graduation. All charts are constructed on the spherical projection, and nearly all are redrawn, engraved and printed upon Mercator's projection. Upon the lakes the change is hardly necessary, but upon larger sheets of water courses are more readily taken from Mercator's than the other.

For the graduation the latitudes and longitudes must be obtained of two points near the extremes of the sheets. The latitudes are obtained as explained; the longitudes must be derived from meridian distances or by the use of signals, the former being the method most usually adopted on Hydrographic Surveys. The errors of the chronometers carried on board the steamer (usually more than three) are ascertained by observation at some point whose longitude is known. As soon as possible after, the error is obtained at the point whose longitude is to be ascertained, and again the error is obtained at the first point. The differences in the errors on the first and third occasions give the rate of the chronometers during the interval, and by interpolation the errors can be found at the first place at the instant the error was observed at the second. The differences give the difference of longitude.

If possible four or five days' work will be taken, as observations can be taken during a day, the move made during the night to the next place, when the following day is used for observations and a return made at night again. By combining observations on the first and third days with those on the second, one difference is obtained; then taking those on second and fourth days with third another difference may be had, and so on.

Knowing the latitudes and longitudes of the extreme stations, the distance in feet between them may be easily calculated and gives a check upon the accuracy of the triangulation.

During the winter the charts are prepared for engraving (which is done at the Admiralty, London),

5

and sailing directions are written giving more details than can be advantageously placed upon the charts.

Observations are carried on during the season for the purpose of ascertaining the magnetic declination at favourable points. For this purpose a Unifilar Magnetometer is used.

The temperature of the water at various depths has been taken in Georgian Bay where the water is deep, but in Lake Erie this has not been attempted on account of the shallow water.

The staff of the Canadian Hydrographic Survey consists of three officers, who are supplied with a suitable steamer especially fitted for the service. The season lasts from May 1st to November 1st, and in that time about 800 square miles of water are carefully sounded, giving about 1,000 nautical miles of ship sounding, and the same of boat sounding, off about eighty miles of shore line.

CHAPTER II.

The Climate of the Dominion.

By Robert F. Stupart, Director of the Meteorological Service of Canada, Dominion Observatory, Toronto.

IN the Dominion of Canada, a country embracing half the continent, we naturally find a very diversified climate ; on the Pacific coast, with the Pacific Ocean on the one side and lofty mountain ranges on the other, it is moist and temperate, while on the east side of the Rocky Mountains, on the high level plateaus of the North-West Territories and in Manitoba, is found a climate with extremes of temperature, but withal a bright, dry, bracing, and healthy atmosphere ; then further east, where we might expect to encounter extremes of heat and cold, such extremes are modified by the influence of the Great Lakes. In the valleys of the St. Lawrence and Ottawa Rivers, a cold but bright bracing winter is followed by a long, warm and delightful summer, while the Maritime Provinces, lying between the same parallels of latitude as France, and with shores laved by the waters of the Atlantic, rejoice in a climate, the praises of which have been sung by successive generations of their people from the old Acadians to the present day.

The climate of Vancouver Island, as of all other parts of British Columbia, varies much with the orographic features of the country. The annual rainfall along the exposed western coast of the island and thence northward to Alaska is very great, generally exceeding one hundred inches. In the south-eastern part of the island between Victoria and Nanaimo the climate does not differ greatly from that found in the north of England; not only does the annual mean temperature agree very closely with that of parts of England, but the mean average of corresponding months is nearly the same. The rainfall is more seasonal in its character. May to September is usually a comparatively dry period, while copious rains fall between October and March.

Crossing the Straits of Georgia to the mainland, we find a warm summer, and a winter increasing in severity as we ascend the Fraser Valley and reach higher levels. At Agassiz, on the Lower Fraser about seventy miles from Vancouver, is situated one of the Dominion Experimental Farms; the average mean temperature of January at this place is $33°$ and of July $64°$, with a daily range of $10°$ in the former month and $26°$ in the latter; the lowest temperature on record is $-13°$ and the highest 97. Frosts seldom occur in May, and there is no record of any during the summer months. The annual rainfall is 67 inches, 66 per cent. of which falls between October 1st and March 31st. This shows very approximately the climate of the Lower Fraser Valley.

Passing to the eastward of the coast ranges the climate changes. The prevalent westerly winds from the Pacific are deprived of much of their moisture on the western side of these ranges and then eastward to the Selkirks, in the Kootenay and Yale Districts, the rainfall and snowfall are nowhere great and at some places are very scant. The annual range of temperature here is large, as is also the daily range during the summer months. Extremely low temperatures are occasionally recorded in winter, and decidedly high temperatures are not infrequent in summer. The Chinook or Foehn effect is very marked in these regions; the moisture laden air from the ocean is deprived of much of its moisture in passing across the coast ranges and is mechanically heated as it descends the eastern slopes,

and thus rendered more susceptible of absorbing moisture and incapable of giving rain; this effect is most
marked under the immediate lee of the mountain
range.

The following table shows the temperature record
at five stations in British Columbia :—

TABLE L

The average mean highest, mean lowest and mean temperature ; the highest and lowest temperature and mean daily range ; also per cent. of cloud and precipitation in inches.

VICTORIA.	JAN.	FEB.	MAR.	APR.	MAY.	JUNE.	JULY.	AUG.	SEPT.	OCT.	NOV.	DEC.	YEAR.
	°	°	°	°	°	°	°	°	°	°	°	°	°
Mean highest......	41·7	43·8	50·1	54·7	61·6	65·3	69·6	69·0	63·3	56·0	48·1	45·7	55·7
" lowest......	32·7	33·6	36·6	39·3	44·1	47·6	49·5	49·5	45·9	42·9	38·2	36·8	41·4
" temperature..	37·2	38·7	43·3	47·0	52·8	56·5	59·6	59·3	54·6	49·5	43·2	41·3	48·6
" daily range.	9·0	10·2	13·5	15·4	17·5	17·7	20·1	19·5	17·4	13·1	9·9	8·9	14·4
Absolute highest ..	56	60	68	74	82	86	89	86	80	70	60	59	89
" lowest....	-1	6	20	24	30	36	37	37	31	22	18	8	-1
Per cent. of cloud ..	77	79	67	70	59	56	43	36	50	67	77	79	64
Precipitation	5·13	3·53	2·57	2·11	1·21	1·16	0·46	0·59	1·98	3·11	6·13	7·53	35·51
AGASSIZ.													
Mean highest......	35·9	43·5	52·0	55·2	62·9	69·5	76·7	79·0	67·4	61·3	44·4	40·8	57·5
" lowest	28·2	31·7	35·9	38·3	44·8	48·3	51·0	50·5	46·7	42·1	34·1	31·9	40·3
" temperature..	33·0	37·6	44·0	46·7	53·9	58·9	63·9	64·7	57·0	51·7	39·3	36·3	48·9
" daily range.	9·7	11·8	16·1	16·9	18·1	21·2	25·7	28·5	20·7	19·2	10·3	8·9	17·3
Absolute highest ..	57	64	74	82	90	95	95	97	90	82	73	58	97
" lowest....	-13	-12	16	28	30	36	38	38	32	29	9	8	-13
Per cent. of cloud ..	62	62	56	60	55	52	35	30	43	55	67	67	54
Precipitation	7·29	6·68	5·47	5·49	4·85	3·97	1·55	1·62	5·25	6·56	8·69	9·43	66·85

KAMLOOPS.

Mean highest	56·9	34·6	39·5	56·8	69·6	83·0	82·9	77·3	68·2	59·8	49·1	33·0	29·6
" lowest	36·0	23·3	27·2	38·6	44·6	53·6	54·6	49·0	45·3	36·7	27·1	15·6	15·0
" temperature	46·5	29·0	33·3	47·7	57·1	68·3	68·8	63·6	56·8	48·2	38·1	24·3	22·3
" daily range	21·0	11·3	12·3	18·2	25·0	29·4	28·3	27·4	22·9	23·1	22·0	17·4	14·6
Absolute highest	101	56	65	82	87	97	101	101	86	75	69	61	56
" lowest	-27	-17	-22	16	31	39	44	39	26	24	-5	-27	-27
Per cent of cloud	59	79	73	52	52	33	47	55	70	62	50	67	72
Precipitation	11·55	0·80	1·00	0·72	1·00	0·53	1·42	1·49	1·45	0·50	0·63	1·26	0·75

MISSION VALLEY.

Mean highest	54·9	33·1	35·5	54·4	63·9	80·3	81·2	74·0	64·9	58·1	45·8	36·6	28·9
" lowest	31·4	21·6	23·3	30·4	37·5	44·8	46·8	42·9	39·6	31·9	23·9	19·8	14·7
" temperature	43·2	27·3	30·4	42·7	50·7	62·5	64·0	58·4	52·2	45·0	34·8	28·2	21·8
" daily range	23·3	11·5	12·2	24·5	26·4	35·5	34·4	31·1	25·3	26·2	21·9	16·8	14·2
Absolute highest	98	47	56	74	82	95	98	94	86	79	65	55	46
" lowest	-17	-12	-17	16	20	31	32	29	25	14	-6	-15	-10
Precipitation	12·28	1·85	1·90	0·70	1·51	0·23	0·38	0·89	1·57	0·64	0·23	1·05	1·33

SALMON ARM.

Mean highest	54·3	34·9	36·7	53·1	62·2	78·7	80·3	72·5	61·2	57·7	46·1	36·9	28·6
" lowest	32·1	24·0	23·3	33·0	37·9	45·9	47·5	44·4	40·0	33·5	24·1	18·6	13·1
" temperature	43·2	29·4	30·0	43·1	50·1	62·3	63·9	58·4	52·1	45·6	35·1	27·7	20·9
" daily range	22·2	10·9	13·4	20·1	24·3	32·8	32·8	28·1	24·2	24·2	22·0	18·3	15·5
Absolute highest	94·3	61·9	63·3	71·3	84·2	94·3	90·8	91·3	88·8	76·3	59·9	58·1	50·4
" lowest	-21	2	-21	21	25	34	37	27	28	23	-1	-20	-11
Precipitation	17·40	2·10	2·13	1·65	1·62	0·43	0·67	1·67	1·48	1·34	0·52	1·28	2·53

The salient features of the climate of the Canadian
North-West Territories are a clear bracing atmosphere
during the greater part of the year, cold winters and
warm summers, and a small rain and snowfall. As
shown by chart No. 2, the mean temperature for
July at Winnipeg is 66°, and at Prince Albert 62°.
The former temperature is higher than in any part of
England, and the latter is very similar to that found
in many parts of the southern counties, the diurnal
range, however, is different from any found in
England, the average daily maximum temperature at
Winnipeg being 78°, with a minimum of 53°, and at
Prince Albert a maximum of 76°, with a minimum of
48°; and owing to these high day temperatures with
much sunshine the crops come to maturity quickly.
In April the 40° mean temperature isotherm passes
through Alberta and Assiniboia, while an average
daily maximum of 58° in Southern Alberta gradually
lowers northward and eastward until 50° is reached at
Prince Albert and near the boundary of Manitoba;
this indicates a spring slightly in advance of that of
south-western Ontario. Spring in this month also
makes rapid strides in Manitoba, where the average
mean temperature is from 35° to 37°, and the daily
maximum about 11° higher. In considering the climate
of the Canadian prairies the fact should not be lost
sight of that although the total annual precipitation
only averages 13·35 inches for the Territories and
17·34 inches for Manitoba, the amounts falling between
April 1st and October 1st, are respectively 9·39 inches
and 12·87 inches, or 70·3 and 74·2 per cent. of the
whole. The average 12·87 inches in Manitoba is not
far short of the average for Ontario during the same
six months. As in Ontario, much of the rain in summer falls during thunder storms. It is rarely, if ever,
that the country is visited by tornadoes such as are
common on the more southern prairies.

In the western portion of Athabasca, near the moun-

tains, signs of spring multiply as April advances, but
in the eastern portion and thence to Hudson Bay,
together with all the northern portions of Keewatin,
winter lingers until quite the end of the month.

The Province of Ontario can boast of as many
distinctly different climates as can any country in the
world. That part of the province which lies immediately north and north-east of Lake Superior, and which
forms the northern watershed of that great lake, has
a long cold winter, and at times extremely low temperatures are recorded ; indeed scarcely a season passes
without 50° below zero being registered at White
River, a station on the Canadian Pacific railway. As
a rule the snow does not disappear from the woods
until the beginning of May, after which time, however,
the summer advances very rapidly, and four months
of superb weather follow. Travelling east and south-
east, the climate quickly improves, and in the valleys
of the Ottawa and the Upper St. Lawrence we find a
moderately cold winter, but a singularly exhilarating,
bracing atmosphere makes even a zero temperature by
no means unpleasant. Signs of spring are not wanting
early in April, and by the beginning of May foliage is
well advanced, and then follows a decidedly warm
summer. The whole of this region is, between the
middle of May and middle of September, included
between the same isotherms as the greater portion of
France, and after a protracted autumn, winter sets in
again before December. The mean annual temperature of Montreal is 41 8, and of St. Petersburg, Russia,
38°7 ; a comparison of the annual curves of the two
places is interesting. The mean for January at Montreal is 5° lower than at St. Petersburg ; in February it
is but 1° lower, and then the Montreal curve rises
steadily above the other until in August it is 6° higher ;
after this the two curves draw together, and by
December are coincident.

In the peninsula of Ontario, or that portion of the

province which lies east of Lake Huron and north of
Lake Erie and the western portion of Lake Ontario,
the winters are by no means severe and the summers
are seldom oppressively hot; this being due to the
tempering influence of the lakes by which this portion
of Ontario is surrounded. In the western counties the
April mean temperature corresponds nearly to that of
southern Scotland, and in May the mean temperature
of the whole district is slightly higher than for the
south of England. The temperature conditions during
the summer months may, as in the Ottawa and St.
Lawrence valleys, be compared with those of France;
the normal temperature for July ranging between 66°
and 72°. September and October are generally delightful months and seldom does snow remain on the
ground until well on in December, except on the high
lands of the interior counties. That portion of Ontario which lies immediately east of the Georgian Bay,
the District of Muskoka, at an elevation of 740 feet
above the sea, abounding in small lakes, possesses a
wonderfully bracing atmosphere which with a very
high percentage of bright sunshine and a pleasant
temperature, has made this region a summer resort
much frequented by people from the cities and towns
further south.

The annual precipitation of the entire province lies
between thirty and forty inches, which is fairly evenly
distributed throughout the year; in summer, however,
the rain generally falls in thunderstorms and cloudy
and wet days are of rare occurrence.

The summers in the south-western part of Quebec
are as warm as in the greater part of Ontario; in July
the 70° isotherm passes not far south of Montreal, the
65° line passes through Quebec city, and most of the
Gaspe peninsula has a mean temperature somewhere
below 60°. The winters throughout the province are
cold, and between December and March the ground has
usually a deep covering of snow.

The importance of the covering of snow during the more severe winter weather cannot be over estimated, as it protects the roots of trees and herbage, and besides this enables those engaged in the lumbering trade in the northern parts of the provinces to get the timber from the bush to the banks of the streams by which the huge logs are floated down to the mills when in the early spring the rapid melting of snow and ice causes a flooding of all water courses.

The opening of spring in the Maritime Provinces is usually a little later than in southern and western Ontario and the North-West Territories, and somewhat earlier than in the Lower St. Lawrence valley ; on the other hand the summer lingers longer, especially in the Annapolis valley. The summers are, as a rule, not quite so warm as in western Canada, great heat being seldom experienced except very occasionally in the inland stations in New Brunswick.

The average annual precipitation of these provinces is between forty and forty-five inches, except along the southern coast line of Nova Scotia where it is nearly ten inches greater.

The following table based on the records of twenty-two years will indicate very closely the climate of the three provinces :—

TABLE II.

The average mean highest, the mean lowest and mean temperature ; the highest and lowest temperature and mean daily range ; also per cent. of cloud and precipitation in inches.

CHARLOTTETOWN, P.E.I.	JAN.	FEB.	MAR.	APR.	MAY.	JUNE.	JULY.	AUG.	SEPT.	OCT.	NOV.	DEC.	YEAR.
Mean highest........	24·0	25·3	31·3	41·4	54·8	66·4	71·6	72·2	63·7	52·8	40·1	29·8	47·8
" lowest........	6·2	7·0	16·0	27·4	37·7	49·0	56·2	57·1	49·7	40·2	28·3	15·9	32·6
" temperature..	15·1	16·1	23·7	34·4	46·3	57·7	63·9	64·7	56·7	46·5	34·2	22·9	40·2
" daily range...	17·8	18·3	15·3	14·0	17·1	17·4	15·4	15·1	14·0	12·6	11·8	13·9	13·9
Absolute highest...	50·0	46·7	52·5	67·9	79·3	85·0	87·7	88·0	82·1	73·9	62·8	52·0	88·0
" lowest....	−26·7	−16·8	−14·2	−2·4	25·8	35·4	42·0	43·5	34·1	26·7	1·3	−17·9	−26·7
Per cent. of cloud..	61	58	64	57	57	58	53	59	52	70	71	71	61
Precipitation	4·00	3·25	3·09	2·61	3·06	2·60	3·43	3·96	3·35	4·65	3·74	3·98	41·78
HALIFAX, N.S.													
Mean highest.......	30·9	31·6	36·5	46·6	58·4	68·2	73·9	74·3	67·6	56·2	44·2	34·3	51·9
" lowest.......	13·1	13·9	20·8	29·9	38·9	47·0	54·4	55·4	48·8	39·8	32·2	19·7	34·5
" temperature..	22·0	22·7	28·7	38·2	48·7	57·6	64·2	64·8	58·2	48·0	38·2	27·0	43·2
" daily range...	17·8	17·7	15·7	16·7	19·5	21·2	19·5	18·9	18·8	16·4	12·0	14·6	17·4
Absolute highest...	54·9	50·0	54·5	76·2	88·0	92·8	93·0	93·1	85·0	79·6	64·5	54·8	93·1
" lowest....	−15·8	−16·9	−9·0	7·2	24·0	32·9	40·8	42·3	32·0	23·4	4·3	−11·2	−16·9
Per cent. of cloud..	62	58	68	59	63	64	57	61	52	63	65	66	61
Precipitation	5·63	4·94	5·15	4·00	4·43	3·63	3·43	3·96	3·53	5·21	5·26	5·52	54·74

FREDERICTON, N.B.

Mean highest	23·3	26·3	35·2	48·9	63·2	66·2	75·9	73·6	65·5	52·3	41·7	27·3	50·5
" lowest	2·8	3·9	16·6	28·1	39·9	49·1	54·4	53·5	44·9	34·4	24·9	9·0	30·1
(6 observ. each day) Mean temperature	11·8	15·8	25·4	37·9	51·1	60·6	66·0	64·1	56·2	44·1	32·0	18·6	40·3
" daily range	20·5	22·4	19·2	20·8	23·3	25·1	21·5	20·1	20·6	17·9	16·8	18·3	20·4
Absolute highest	51·8	50·8	64·7	76·8	91·7	96·5	96·2	94·7	87·7	82·1	63·7	57·7	96·5
" lowest	-34·5	30·5	27·2	3·6	21·4	32·0	38·0	39·0	24·9	14·8	-15·7	-31·5	-34·5
Per cent. of cloud	57	53	56	54	58	56	52	53	56	60	62	52	56
Precipitation	2·43	3·76	4·12	2·59	4·23	3·64	3·79	4·18	3·21	3·93	4·21	3·62	43·71

The Great Lakes never freeze over ; but usually most of the harbours are closed with ice by about the middle of December, and remain frozen over until the end of March or beginning of April. The average date of the closing of navigation on the St. Lawrence River at Montreal is December 16th, and of the opening, April 21st. Harbours in the Gulf of St. Lawrence are like-wise closed by ice during the winter months, but on the Bay of Fundy and coast of Nova Scotia they are open all the year round.

A very casual inspection of the following table (III.) showing the normal percentage of bright sun-shine in various parts of the Dominion as registered by Campbell-Stokes recorders will render it very obvious that Canada is a country of clear skies. There are few if any places in England that have a larger normal annual percentage than 36 and there are many as low as 25, whereas in Canada most stations exceed 40 and some few have as high a percentage as 46. In England at but few places does the normal of any summer month exceed 45 per cent. At German sta-tions the August maximum averages under 50 per cent. and in a few cases reaches 52. In the south much higher values are obtained, Vienna 54, Zurich 57, Trieste 66, Lugano 67, Rome 75, Madrid 84. These figures show that it is only the southern parts of Europe that have more sunshine in the summer months than Canada.

References :—" The Reports of the Dominion Me-teorological Service ;" "Geological Survey Reports ;" "Report of Cruise of H. M. S. Challenger, Vol. II, Part V.;" "Reports of the Meteorological Office, Lon-don."

TABLE III

Percentage of bright sunshine at fourteen stations in the Dominion (100 being constant sunshine).

	JAN.	FEB.	MAR.	APR.	MAY.	JUNE.	JULY.	AUG.	SEPT.	OCT.	NOV.	DEC.	YEAR.
	%	%	%	%	%	%	%	%	%	%	%	%	%
Esquimalt, B.C.	20	29	35	32	40	43	53	56	36	37	20	16	34
Agassiz, B.C.	21	21	24	21	29	35	46	48	27	30	18	14	28
Battleford, N.W.T.	29	45	46	47	35	58	59	52	31	49	26	31	42
Winnipeg, Man	43	48	51	47	53	54	58	59	34	37	35	37	46
Brandon, "	41	45	47	40	44	46	58	61	47	36	29	32	44
Lindsay, Ont	26	34	44	50	47	56	57	53	53	39	24	21	42
Toronto, "	29	34	43	49	48	56	60	58	56	42	28	22	44
Stratford, "	25	24	36	40	44	54	59	55	45	37	24	17	38
Windsor, "	24	27	39	45	47	55	60	54	48	42	30	20	41
St. Catharines, Ont.	17	23	30	40	46	54	56	57	51	33	21	16	37
Montreal, Que.	33	41	47	52	51	54	59	59	54	40	29	30	46
Fredericton, N.B.	40	44	41	47	46	48	51	50	48	43	33	36	44
Sydney, N.S.	31	34	32	39	43	47	48	46	49	34	27	19	37

CHAPTER III.

Sketch of Canadian Zoology.

Prof. Ramsay Wright, University of Toronto.

IT seems desirable within the limits allotted to this sketch to remind the naturalist who has not previously visited Canada of the interesting forms of animal life which he may meet with, and to indicate some sources of information* as to these rather than to attempt a survey of the Canadian fauna from a zoogeographic standpoint, especially as the data for the latter are still very incomplete.

It may be premised, however, that recent students of the distribution of life on the North American continent concur in recognizing two provinces, a northern and a southern, the greater portion of the former being situated within Canadian Territory, of the latter within the United States. Dr. Merriam+ distinguishes these as Boreal and Austral respectively, and observes, apropos of his conclusion that the southward distribution of northern forms is determined by the mean temperature of the hottest part of the year, that the southern boundary of the Boreal province corresponds to the isotherm of 18° C (64·4 F) for the six hottest consecutive weeks. This is not substantially different from the isotherm of 65 F for July laid down on Mr. Stupart's Chart, No. 2, and it therefore follows that the most northerly (or transition) zone of the Austral

---

*Jordan's Manual of the Vertebrates of the Northern United States, Chicago, 1888, is also very convenient for the student of Canadian Zoology.

+Nat. Geog. Mag. VI.

province advances beyond the international boundary
line in three regions, viz., in Ontario as far north as
the Georgian Bay, in the western plains chiefly in Assiniboïa and in British Columbia. In these regions
consequently there is a certain infusion of southern
types into the Canadian fauna.

There have further been recognized within the Boreal
province three zones, Arctic, Hudsonian and Canadian,
and Dr. Merriam is inclined to assign the isotherms of
10° C (50° F) and 14° C (57° F), as the southern boundaries of the first two, and, as has already been indicated,
18° C (64°·4° F) as that of the last. It has been ascertained that the distribution of the mammals agrees
fairly well with the temperature zones thus marked
out, but the data as to other groups are insufficient.

Mr. J. B. Tyrrell* has published a convenient list of
the mammals of Canada, in which he has incorporated a
good deal of information as to the range of various
species accumulated by himself and other members of
the Geological Survey, as well as references to the literature of the subject.

The list contains over 120 species, more than a
third of which are rodents, as one would expect from
the great development of this order in North America.
Among these are some of the commonest forms of
mammals which the visitor is likely to make acquain-
tance with. Perhaps the little red squirrel (*Sciurus
hudsonicus*) and the striped chipmunk (*Tamias stria-
tus*) will first attract his attention, but he may also
meet with the extremes of diversity of habit within
the Sciuridæ in the flying squirrel (*Sciuropterus vol-
ans*) which is widely distributed, and the Prairie
ground-squirrels (*Spermophilus*) of Manitoba and the
western plains. If he has observed the Alpine mar-
mot in some lonely Swiss pass, he will hardly recog-

---

*Trans. Can. Inst., Toronto, 1888.

6

nize its congener, the solitary woodchuck (*Arctomys monax*) which is everywhere common in the wooded banks of streams. The larger species from the northern Rocky Mountains (*A. caligatus*), on the other hand, appears to resemble the European form more closely in its habits.

A very characteristic rodent of the western plains is the northern pocket gopher, a member of the singular family Geomyidæ. This, as Dr. Merriam has shown, is a group of Mexican origin, which has pushed up into British Columbia and the plains of the Saskatchewan, being represented there by the species *Thomomys talpoides*. They are subterranean creatures, forming mounds over their burrows and carrying off the surplus roots and tubers on which they feed in the cheek pouches from which they receive their name.

Of the aquatic rodents the musk-rat (*Fiber zibethicus*) is everywhere common, but the beaver is no longer to be met with in the vicinity of civilization, although the preservation of Algonquin Park appears to have resulted in increasing their number there.

Among the larger members of the order is the Canada porcupine (*Erethizon dorsatus*), which extends northwards to the limit of trees, and is still common in the less settled districts.

Lastly, the Leporidæ are well represented, the commonest forms being the wood-hare or cotton-tail (*Lepus sylvaticus*) the varying hare (*L. americanus*), and the jack-rabbit of the western plains (*L. campestris*).

The Carnivora contribute almost a quarter of the mammalian fauna, and the chief proportion of the valuable furs, such as those of the mink (*Putorius vison*), the skunk (*Mephitis mephitica*), and the otter (*Lutra hudsonica*). The Insectivora likewise are numerous, including some twenty species, of which the moles are particularly interesting on account of the forms themselves and their geographical distribution.* The same remark applies to the bats,† of which six species occur in Canada.

Of special interest to the sportsman are the various species of Ungulates. The characteristic genus of the Cervidæ is Cariacus, represented by the common and widely distributed Virginia deer (*C. virginianus*), and in the west, by a nearly allied form (*C. leucurus*), as well as by the mule deer (*C. macrotis*), marked by its long ears and dark dorsal stripe. The remaining genera, Cervus, Alce and Rangifer, are all European as well. To the first genus belongs *C. canadensis*, the great Canada stag, erroneously called the elk in America, and according to Mr. Tyrrell not the wapiti (as it is now very generally styled), but the waskasew of the Indians. The true elk or moose (*Alce alces americanus*), and the caribou (*Rangifer tarandus caribou* and *grœn-landicus*), are probably only varieties of the European elk and reindeer.

The moose, which had been becoming scarce in the northern forests of Ontario, has been considerably commoner since a five years' protection was afforded to it. It is now also pushing its way further towards the north-west boundary of the Dominion.

The two sub-species of reindeer are known as the woodland and barren-ground caribou, the latter confined to the treeless districts west of Hudson Bay, the former extending from Newfoundland to the Pacific coast.

Our species of *Bovidæ* are decidedly less accessible. They include the prong-horn (*Antilocapra ameri-cana*), which is in reality a southern form which pushes northward on the western plains, and the two characteristic forms of the northern Rocky Mountains

---

* True. Revision of the American Moles, Proc. U.S. Nat. Mus. Vol. XIX.

† H. Allen. Monograph of the Bats of North America. Bull. U.S. Nat. Mus., No. 43.

—the mountain sheep and goat (*Ovis montana* and *Haplocerus montanus*). Finally, the musk-ox (*Ovibos moschatus*), the only exclusively American of the Arctic mammalia, still abounds in the barren grounds, while the bison (*Bos americanus*) is practically exterminated, although a few "wood buffalo" are supposed to exist near the head waters of the Mackenzie River.

It is natural that Ornithology should have received more attention than Mammalogy, and that its literature should therefore be more extensive. Mr. Montague Chamberlain's catalogue of Canadian birds* has for its scope the whole Dominion, and the same author's edition of Nuttall's Ornithology† devotes special attention to Canadian forms. There exist, however, special faunistic works of great interest for the more limited regions with which they deal. Such are the Birds of Ontario, by Thomas McIlwraith,‡ Les Oiseaux de Quebec,§ by C. E. Dionne; and the Birds of Manitoba, by Ernest E. Thompson.¶ Professor Macoun has devoted special attention to the avian fauna of the great North-west, and may be expected to give some of the results of his wide experience at the B. A. meeting.

Ornithologists who are especially interested in oology may be glad to know of the extensive collections of Mr. Walter Raine of Toronto, who has published an account of his "Bird Nesting in North-West Canada."||

---

*Saint John, N.B., 1887.
†Boston : Little, Brown & Co., 1891.
‡Hamilton, Ont.
§Quebec, 1889.
¶Proc. U.S. Nat. Mus. Vol. XIII., 1891.
|Toronto : Hunter, Rose & Co., 1892.

The reptiles occurring within the Dominion chiefly belong to the Ophidia and Chelonia, a few lizards of the genera Eumeces and Sceloporus extending beyond the international boundary line in the west. Some fifteen species of Colubridæ occur in the east, for which alone data are obtainable, and the commonest of these are, in the order in which they are described in Professor Cope's recent revision,* the milk-snake (*Ophibolus doliatus triangulus*), the ring-neck (*Diadophis punctatus*), the grass snake (*Liopeltis vernalis*), the racer (*Bascanium constrictor*), the fox-snake (*Coluber vulpinus*), the pig-nosed snake (*Heterodon platyrhinus*), various garter snakes (*Eutania*, especially *E. sirtalis*), various species of Natrix and Storeria (*N. leberis* and *rigida*, *N. fasciata sipedon*, *S. dekayi* and *occipitomaculata.*)

Two species, finally, of rattlesnakes occur, but their area of distribution is much more limited than formerly : these are *Crotalus horridus*, and *Crotalophorus catenatus tergeminus.*

Four families of Chelonia are represented in the Dominion ; the Emydidæ by the common pond-turtle (*Chrysemys picta*), as well as by the less common spotted turtle (*Clemmys guttata*), geographic turtle (*Graptemys geographica*), and box-turtle (*Cistudo (Terrapene) carolina*); the Chelydridæ by the common snapping turtle (*Chelydra serpentina*), the Kinosternidæ by the musk-turtle (*Aromochelys odorata*), and the Tryonychidæ by the more southerly soft-shelled turtle (*Platypeltis spinifer*).

Canada contrasts very strongly with Great Britain in its abundant Amphibian Fauna. One of the most interesting members thereof, the Menobranch or Mud-Puppy (*Necturus maculatus Raf.*), is common in the

---

* A critical review of the characters and variations of the Snakes of North America: Washington: Proc. U. S. Nat. Mus., vol. XIV,

eastern portion of the Dominion, while the British newts
are represented by a large series of salamanders some
of them characteristic of the Atlantic, others confined
to the Pacific slope ; of the former the commonest are
the little spotted newts (*Diemyctylus viridescens Raf.*),
but species of Amblystoma, Plethodon and Desmo-
gnathus are widely distributed.   One of the Ambly-
stomas common in the western plains (*A. tigrinum*)
rivals the Mexican Axolotl in the size which it frequently reaches before undergoing its metamorphosis.

In spring and early summer the lakes and ponds are
rendered vocal by the numerous species of frogs and
toads, as many as eight distinct species of Rana and
four or five Hylidæ contributing to the concert.

From an ichthyological standpoint the Dominion
presents many interesting features, especially in the
great development of certain families of physostomous
Teleosts, which the vast inland seas have favoured.

The Siluroids, represented in Europe only by the
"wels" of the Danube, here furnish some of our commonest species like the small horned pout (*Ameiurus
nebulosus*) and the great forktailed catfish of the lakes
(*A. lacustris*), which may weigh over 100 lb.   Again,
the carps are replaced by a curious series of suckers or
catostomids, which in spring ascend the rivers in vast
numbers, but are of little use for food purposes.

The herring and salmon families exhibit a tendency
to develop land-locked forms.   Completely fresh
water clupeids are the hyodons or moon-eyes and
golden eyes of the great lakes and the North-west,
while the shad (*Alosa sapidissima*) and alewife
(*Pomolobus pseudoharengus*) penetrate from the coast
far into the interior and may in the case of the latter
be completely land-locked.

Like the large European bodies of fresh water, our
lakes contain certain characteristic coregonids, which,

like their Swiss congeners, offer comparatively unstable specific characters.

One of the commonest, as well as the most valuable commercially, is the whitefish (*C. clupeiformis*) of the great lakes, which gives place to the Tullibee in the Lake of the Woods and Manitoba, and in the North-west to less known species like the Inconnu.

From both the Atlantic and Pacific, salmon penetrate far into the interior at spawning time. The Atlantic salmon (*S. salar*) were at one time abundant in Lake Ontario and have given rise to a land-locked variety in Lake St. John, Quebec, the Ouananiche, which is much sought after by anglers. Almost incredible are the accounts of the vast numbers of salmon (*Oncorhynchus nerka* and *quinnat*) which ascend the rivers of the Pacific coast, from spring till autumn, hurrying up in certain cases as much as a thousand miles before spawning.

Among the most important fish commercially is the lake trout (*Salvelinus (Cristivomer) namaycush*), a charr which in the Great Lakes may attain the weight of 100 lb., and is very variable in its colouration and the size which it reaches. The same is true of the brook trout (*Salvelinus fontinalis*), another charr which may attain a large size in such bodies of water as Lake Nipigon, or in tidal rivers where it is called the sea trout, but in smaller lakes and streams rarely exceeds one or two pounds.

Another example of the influence of large bodies of water in leading to increased size of their inmates, is perhaps to be found in the pike family. This is represented in Canada by the common European form *Esox lucius*, but also by a much larger form, the maskinonge (*Esox masquinongy*), which furnishes excellent sport where it occurs, and has been recorded as attaining 100 lb. in weight.

Among the hard-rayed or physoclyst fishes, the most characteristic is the bass family (*Centrarchidæ*), which

yield not only the large and small mouthed black bass
(*Micropterus dolomieu* and *salmoides*), but also the
smaller sunfish and rock-bass.

The European pike-perches, or sanders, are represented by two species which attain considerable commercial importance, and are known as pickerel or doré
(*Stizostedion vitreum* and *canadense*), the perches proper by the common yellow perch (*P. flavescens*), and by
a host of brilliantly coloured darters, and the marine
Sciænids by the Lake Huron drum (*Haploidonotus*).
Certain forms of the allied marine Serranidæ, the white
and striped bass (*Roccus chrysops* and *lineatus*), are
valued as food fishes. The former occurs in the Great
Lakes, the latter ascends the rivers of the Atlantic
coast for spawning purposes, and may penetrate far
inland.

Most characteristic for the inland waters are the
three genera of ganoids, Acipenser, Lepidosteus and
Amia. The lake sturgeon (*Acipenser rubicundus*), is
common in Lake Erie and the Upper Lakes, while the
garpike (*Lepidosteus osseus*), and the bowfin (*Amia
calva*), are less abundant, but widely distributed.

No attempt has been made to touch upon the rich
marine vertebrate fauna of the Dominion, nor is it
possible to do more than refer to the investigations of
Sir J. W. Dawson and Mr. Whiteaves on the inverte-
brate fauna of the Gulf of St. Lawrence—perhaps the
best explored part of our coasts.

Along with Ornithology, Entomology has claimed a
large number of students. One of the most important
serials devoted to this branch of American Zoology is
the *Canadian Entomologist*, the organ of the Entomological Society of Ontario, which has appeared regularly for the last thirty years, and contains many
valuable papers on the insect fauna of Canada. The

society, which has its headquarters at London, Ont., owns the most complete collection of Canadian insects.

Other divisions of Zoology have likewise had their devotees, as a glance through the proceedings of the various learned societies, such as the Canadian Institute of Toronto, the Natural History Societies of Montreal, Halifax and Victoria, the Field Naturalists' Club of Ottawa, etc., serves to show.

Perhaps less attention has been given to the lake faunas than might have been expected in a country so intersected everywhere by small lakes, besides being part possessor of the largest bodies of fresh water in the world. But the interest in these, recently awakened by the establishment of fresh-water stations in various parts of Europe and America, may be expected to lead to more comprehensive studies than have yet been attempted.

It may be well to state for the convenience of the sportsman, that the regulations affecting game in the various provinces of Canada and in the various states of the Union will be found in a convenient form in the Book of the Game Laws, published by the Forest and Stream Publishing Co., New York City.

CHAPTER IV.

Sketch of the Flora of Canada.

By Professor John Macoun, Naturalist to the Geological Survey of Canada.

IN a general sketch of the flora of the Dominion of Canada, the whole northern portion of the North American continent must be considered, including Newfoundland on the east and Alaska on the west. This immense region, extending from Cape Race, the most easterly point of Newfoundland, to Behring Straits on the west, is in round numbers 3,500 miles wide. On the south, the forty-ninth parallel forms the boundary from the Pacific Ocean eastward to the Lake of the Woods, from thence to where it cuts the forty-fifth parallel, it follows a tributary of Lake Superior, the great lakes, and the St. Lawrence River itself. The northern boundary of New York, New Hampshire, Vermont and Maine, form the southern boundary to the sea at St. Stephen, New Brunswick.

The chief features of the northern and eastern sections extending westerly to the Mackenzie River, are its plains, lakes, rivers and forests, and the paucity of its flora as regards species, the greater number of which are identical with those of northern Europe, or very closely related to them. The south-western or prairie region has a flora which is quite distinct both in origin and appearance from that of the forest region to the north and east. South-western Ontario has a flora that in greater part has a southern origin, and which in very many respects differs from that of all the other parts of the Dominion, and includes many species of shrubs and trees that do not grow naturally outside of its limits.

The whole of the Dominion east of the Rocky Mountains may be called a plain, as it rises at no point into anything that could be called a chain of mountains. The only chain of heights are the Laurentides, extending up the St. Lawrence and along the Georgian Bay and Lake Superior. West of Quebec city to the Rocky Mountains there is no point above 2,000 feet until the high plains become an elevated plateau, but altogether destitute of mountains. The source of the St. Lawrence (Lake Nipigon), 1,900 miles from the sea, is less than 800 feet above tide water.

The Rocky Mountains, extending in a north-westerly direction from latitude 49° to the Arctic Sea, are both a barrier to the western extension of the prairie flora and a means of extending the distribution of the Arctic, for many species found on the Arctic coast are found in the Rockies at aititudes ranging from 7,000 to 9,000 feet.

British Columbia consists of a series of mountains, plateaus and valleys, that have a very varied herbaceous vegetation, and as a consequence we have on the mountain summits an Arctic flora with a marked change to Alaskan species as we ascend the Coast Range and the mountains on Vancouver Island. On the dry region about Kamloops, Okanagan, and Spence's Bridge, there are many species that have their home to the south in the dry districts of Washington. On the other hand, the coast flora, and especially that of the vicinity of Victoria, has much in common with northern California and Oregon.

From the foregoing it may be seen that our flora is made up of a series of fragments that have had, each, a different origin ; the more northerly and high mountain species being circumpolar or derivatives from those of northern Europe and northern Asia. The species in the coniferous and poplar forests are also of

northern origin, but those in the deciduous-leaved
forests of the eastern provinces and Ontario are
undoubtedly characteristic of America and have a
much greater development to the south. Genera that
are characteristic of these forests are *Desmodium,
Uvularia, Trillium, Podophyllum, Hydrastis, Phlox,
Dicentra, Sanguinaria, Medeola*, and many others.

In the prairie provinces the species are of a south-
westerly origin, though many are identical with the
mountain species of that region of the United States.
Eastern species of herbaceous and woody plants extend
far to the west in the stream valleys and wooded
ravines, and do not finally disappear until the more arid
districts are reached. In the foot hills of the Rocky
Mountains many western species find a home, and they
too descend to the plains and spread themselves eastward until stopped by the light rainfall of the prairie.

The European botanist when first landing on our
shores or entering the country by any of the United
States railways will be struck by the similarity between
the plants he meets with and those of his own country.
This seeming resemblance only extends to the roadsides and cultivated grounds. What he sees are
immigrants, and it is only in the forest he will see indi-
genous plants. In trying to get a knowledge of the
native flora no person should collect anything along
the roadsides or in cultivated fields, because not ten
per cent. of the species he sees are natives. Our native
species seldom become weeds, as they were chiefly forest
species, and with the forest many of them disappear.

Lying between Hudson Bay and the Gulf of St. Lawrence is the extensive tract named Labrador, the
interior of which until lately was quite unknown. The
area of this peninsula is over 500,000 square miles, and
the paucity of its flora may be learned when it is
known that the flowering plants and ferns that occur in
it number less than 1,000 species. It is only on the
coasts and the more elevated mountains that the true

Arctic flora is found, and even this only in the north-eastern part where the Arctic ice is forced on shore. The characteristic Arctic species found here are *Ranunculus pygmœus*, Wahl, and *R. nivalis* L.; *Papaver nudicaulis*, L.; *Draba alpina*, L., *D. stellata*, Jacq. and *D. aurea*, Wahl.; *Silene acaulis*, L., *Lychnis alpina*, L. and *L. apetala*, L.; *Potentilla maculata*, Poir.; *Saxifraga oppositifolia*, L., *S. rivularis*, L., *S. cernua*, L. and *S. nivalis*, L.; *Sedum rhodiola*, Db.; *Erigeron uniflorum*, L., *Antennaria alpina*, Gœrtn.; *Campanula uniflora*, L.; *Ledum palustre*, L.; *Rhododendron Lapponicum*, Wahl.; *Diapensia Lapponica*. L.; *Pedicularis Lapponica*, L. *P. hirsuta*, L., and *P. flammea*, L. Numerous willows, sedges and grasses which are Arctic or mountainous in their general distribution are to be met with, but the bulk of the flora is identical with that of the sub-arctic or boreal zone of the forest belt that extends to the Mackenzie River. None of the enumerated plants have been observed in the interior of Canada, but all with one or two exceptions are to be found near the snow line in the Rocky Mountains.

Prince Edward Island has nothing peculiar about its vegetation except that both the seaweeds around its shores and its land flora indicate greater warmth in its coast waters than we find on the coasts of Nova Scotia. There is a marked absence of species indicating a boreal or frosty summer climate, while there are undoubted indications of a moist and cool one. One summer spent on the island revealed very little of botanical interest, but showed that Prince Edward Island was climatically the "Green Isle" of the Dominion.

Owing to the position of Nova Scotia it has more the characteristics of an island than a continental mass, and hence a number of species are found there and on the coast of Newfoundland that are never met inland. The general flora, however, is seen to be in general the same as that of the Provinces of New Brunswick, Quebec, and the greater part of Ontario.

A few notable species are *Calluna vulgaris*, Salisb, *Alchemilla vulgaris*, L., *Rhododendron, maximum*, L., *Ilex glabra*, Gray, *Hudsonia ericoides*, L., *Gaylussacia dumosa*, T. & G., and *Schizœa pusilla*, Pursh.

Passing to New Brunswick we find a marked change in the flora, which now takes on a more exclusively American facies, and as we pass westward this becomes more marked until scarcely a trace of the European flora can be detected except on the higher summits. Gradually the eastern species drop out and are replaced by immigrants from the south or the advance guard of the western flora. In the deciduous-leaved forest many species are found that are rare or absent in Nova Scotia but which are common in Western Quebec and Ontario.

Owing to the position of Quebec its flora varies greatly, for while on the shores of the Gulf and the lower reaches of the St. Lawrence many Arctic and sub-arctic species may be found, the conditions have so changed when Quebec city is reached that the Wild Grape (*Vitis riparia*) and the Silver Berry (*Elæagnus argentea*) grow luxuriantly on the Isle of Orleans, and the valley of the St. Lawrence westward shows a constantly increasing ratio of southern forms. Along the shores of the lower part of the river the writer has collected *Thalictrum alpinum*, L., *Vesicaria arctica*, Richards., *Cerastium alpinum*, L., *Arabis alpina*, L., *Saxifraga cœspitosa*, L., and *S. oppositifolia*, L.; and on Mount Albert, one of the Shickshock Mountains, *Silene acaulis*, L., *Lychnis alpina*, L., *Rhododendron Lapponicum*, Wahl., *Cassiope hypnoides*, Don., and many others. On the summit of this mountain at an altitude of 4,000 feet were collected *Vaccinium ovalifolium*, Smith, *Galium Kamtschaticum*, Steller., *Pellœa densa*, Hook, and *Aspidium aculeatum*, Swartz. var. *scopulinum* D. C. Eaton. The two latter have no other known stations east of the Pacific Coast Range and the other two are western species.

Montreal Mountain, on the other hand, may be said to be an eastern extension of the southern flora, as here we have the first assemblage of the representative Ontario flora.

No other province of the Dominion has such a diversified flora as Ontario, caused by the great influx of southern forms in the south-western peninsula bordering on Lake Erie, and the extension of the province westward to Manitoba and northward to James Bay.

To speak in general terms, that part of Ontario north of the Canadian Pacific railway and north and west of Lake Superior, has a flora in no respect different from that of the boreal sections of Quebec and the Maritime Provinces. Along the Ottawa and the St. Lawrence, from Montreal westward, the country gradually improves in climate, and corresponding to this change the flora takes a more southern aspect and trees, shrubs, and all herbaceous plants not hitherto seen become common. In the vicinity of Toronto a marked change takes place and Scarboro' Heights and the Humber Plains seem to be the gathering ground for many species that do not occur in a wild state farther to the east.

Yonge street, which was the great northern highway 100 years ago, is still a divisional point for various reasons, but in none more so than in a botanical sense. West and south of this line a new forest with new shrubs and herbaceous plants meets the eye of the botanist and tells him with unerring certainty that he has entered on a new field for his labors, and if he be a practical man he will soon see that the capabilities of the country increase with the change. Any botanist desirous of collecting many rare Canadian plants in a small space must not fail to visit the Humber Plains and High Park, Toronto, Queenston Heights, the Niagara River and Falls, Hamilton and the district in that vicinity, and any other localities from Point

Edward at the foot of Lake Huron to Fort Erie at the head of the Niagara River. All points are interesting to the botanist, but none more so than from Kingsville to Sarnia, taking in Pelee Island, where vineyards rivaling those of Europe are seen in perfection. Amherstburgh, Windsor, Chatham and Sarnia are easily accessible, and at all these places rare and beautiful species can be obtained.

Should the general flora of the northern forest be desired or the water-plants (Potamogeton) of the country there is no other place so advantageously situated as the Muskoka Lake district, where the diversified scenery of lake, river, rock, and forest-clad promontories will delight the heart of any one, and where botanists of all grades can load themselves with treasures by very little effort, and at the same time suffer neither from fatigue nor lack of first class hotel accommodation.

While the shores of Lake Erie are clothed with vegetation that needs a high winter temperature, the east and north coasts of Lake Superior have a boreal vegetation that shows that the summer temperature of this great lake is quite low. It was the boreal species along the cliffs and near the water that led the early travellers and Agassiz to carry away such erroneous impressions of the Arctic climate of the Lake Superior region; a region which we now know is not climatically unsuited to agriculture. It may not be uninteresting to know that the Great Lakes have, with the exception of Lake Superior, a much earlier growth in spring on the north shores than they have on the south.

Passing out of the forest region, we enter on the vast expanse of natural meadows which constitute the prairie region of the travellers and the Provinces of Manitoba, Assiniboïa, Alberta and part of Saskatchewan. The eastern border is about thirty miles east of Winnipeg, and the western border, the foothills of the

Rocky Mountains 900 miles to the west. This vast region has in many respects a flora quite different from that of the east, north or west in which species of the forest zone predominate. As mentioned in another place, the eastern flora extends westerly in ravines and river bottoms for 150 or 200 miles, but finally disappears, and the true prairie flora is found everywhere except in a few localities where the conditions are favorable to the growth of a few moisture-loving herbaceous species of the forest region.

The advance of northern forest species is checked by the encroachment of the prairie caused by fires in former years, and the intermingling of species peculiar to prairie and forest is well shown in the district between Prince Albert and Edmonton. On the west the advance of prairie species on the eastern slopes and foothills of the Rocky Mountains is no less marked, and the day is not far distant when the whole eastern slope and many interior valleys will be given up to pasturage and the growth of hay for the immense herds and flocks that will feed in summer on the higher slopes and find food and shelter in winter in the valleys.

Much has been spoken and written about the nutritive quality of the grasses of the foothills in Alberta, but the same may be said of the whole prairie region. The same species are common over nearly the whole area, and indeed the only coarse grasses of the dry prairie *Festuca ovina*, L., and *F. scabrella*, Torr., have their greatest development in the foothills where they, with certain species of *Danthonia*, are cut in large quantities for hay. Parts of six seasons spent on the prairie, collecting natural history specimens, give as the grasses of the prairie no less than forty-two genera and one hundred and fifty-six species. Of *Agropyrum, Elymus, Stipa, Bromus, Agrostis, Calamagrostis*, and *Poa*, the best hay and pasture grasses, there are fifty-nine species, so that without the aid of cultivated or

7

foreign species, with the aid of irrigation, we can have hay and pasturage for all purposes. The genus *Carex* furnishes much of the summer food of the native ponies, and one species *C. aristata*, R. Br. has always been their summer food when Indians and half-breeds were on the march. Besides the grasses the prairie produces many Leguminous plants that are valuable for pasture, especially of the genera *Astragalus*, *Vicia*, (Wild Vetch), *Lathyrus* (Wild Pea), of which we have twenty-eight species. The Rose Family is well represented and many species of *Prunus*, *Fragaria*, *Rosa*, *Rubus*, and *Amelanchier* produce fruits which serve as food for both birds and men.

There are a few species found in and around water that are worthy of a passing glance. Old Wives' Lake is saline, and in its waters we find *Ruppia maritima*, L., and along its shores *Heliotropium Curassavicum*, L., both natives of the Atlantic coast, and numerous species of *Chenopodium*, *Atriplex*, and allied genera. On an island in the same lake we find breeding the Ring Billed Gull that winters on the Atlantic coast. In boggy ground near Crane Lake a species of *Downingia*, is found in profusion. If not new, it has no relatives nearer than California. In the same bog the Californian Grebe was breeding in numbers. Still more extraordinary, on Sheep Mountain, close to Waterton Lake near lat. 49°, at an altitude of 7,500 feet, was gathered a mountain poppy which, when submitted to experts at Kew and Washington, was pronounced to be *Papaver Pyrenaicum*. How did it get there ?

Leaving the prairie let us turn eastward to the Atlantic coast and follow the forest belt from lat. 46° north-westerly to where lat. 54° strikes the Rocky Mountains, and we will find a flora that does not vary ten per cent. in the species that inhabit either forest, swamp, lake or stream. In this distance of 2,500 miles the hygrometric conditions seem the same and the apparently severer winter of the west is offset by the universal covering of snow. It might be as well to remark here that accurate meteorological data have shown that Edmonton, in northern Alberta, in lat. 53° 30', has almost the winter climate of Ottawa in lat. 45° 25'.

It is then no fiction to state, as I did in 1872, that the climate of the wooded portion of the North-west is very much like that of northern Ontario. As time passes and this forest belt gets broken up and drained it will be found to be subject to less extremes of cold, heat and drought than the prairie to the south, and the term "fertile belt" will be again applied to the banks of the Saskatchewan as it was in former years.

The mountain region may be said to include all British Columbia, and may be described generally as a high mountain plateau studded with ranges of isolated peaks.

The eastern ranges are included in the term Rocky Mountains, and in these we have most of our higher summits. The vegetation of the elevated prairie (alt. 4000) near the eastern base or foothills of the Rocky Mountains is exceedingly rich and consists of a very varied and most luxuriant growth of herbaceous plants, including a number of rare and interesting Umbellifers. As the slopes are ascended the species of the plain gradually disappear and at 6,000 feet many boreal plants show themselves, and as greater heights are attained the vegetation becomes more Arctic, so that from 7,500-9,000 feet in the Rocky Mountains nearly all the species are either identical or closely related to those found on the "barren grounds" and along the Arctic coast east of Mackenzie River. To obtain a fair knowledge of the flora of the Rocky Mountains a few days' collecting at Banff, in the mountains around Devil's Lake and at Lake Louise, is all that is necessary. At Banff there is a local herbarium in which there is a complete representation of the mountain

flora and with its aid and the ascents of a few mountains which can be done without much labour, considerable knowledge of the mountain flora can be acquired
in a few days.

The Peace River vegetation differs very little from
that of Quebec and the northern prairies, and as far
north as lat. 61° these species predominate and apparently all the country needs is drainage to give it a
climate suitable for all kinds of crops.

The western slopes of the Rocky Mountains begin
to show a mixed flora and both herbaceous and woody
growths have a noticeable increase of western forms.
Both the valley of the Columbia River and the
mountain sides bordering it show by their flora that
we have passed from a comparatively dry climate
into a damp one, and the corresponding change in
both the flora and avian fauna becomes apparent. Any
one now entering the woods along the river or up the
slopes will not fail to notice the thick carpet of moss
and the general dampness, and at the summit of the
Selkirks he will learn that the average snowfall is not
less than thirty feet.

This fact will account at once for the great number
of glaciers at comparatively low altitudes in the Selkirks, and their total absence in the Rockies below
8,500 feet. Owing to the humidity of the atmosphere,
the flora of the Selkirks differs greatly from that of
the Rocky Mountains, and has much in common with
that of the Pacific Coast towards Alaska. Collections
made at Banff, at an altitude of 7,000 feet, will be
quite different from those made at the same altitude
at Glacier in the Selkirks.

West of the "Great Bend" of the Columbia River,
British Columbia becomes a high plateau studded
with mountains and cut into deep narrow valleys. In
some instances, as in the case of the Okanagan Valley,
this plateau has an outlet to the south, and it has
therefore a flora which in part is peculiar to the

American desert, and such species as *Purshia triden-tata*, D.C., and *Artemisia tridentata*, Nutt., and species of *Gilia*, *Aster* and *Erigonum* are found that are met with nowhere else in Canada. The reptilian and avian faunas partake of the same character, and rattle-snakes and lizards with rare southern birds are quite common.

Owing to variations in altitude and the direction of the prevailing winds, British Columbia varies from the aridity of the region just spoken of to the almost con-stant rains on the coast, and while in the Fraser River valley below Yale the vegetation partakes of the char-acter of the tropics in the same valley, fifty miles above Yale, at Lytton, aridity and an almost total absence of rain give almost the same flora as we found about the southern end of Lake Okanagan. That part of British Columbia south of the Canadian Pacific Railway is an elevated plateau which is studded with mountains. In the valleys between and up the mountain sides to an altitude of from 3,000 to 3,500 feet there is little timber and the country is open and covered more or less with grass and a few shrubs. At 3,000 feet trees begin to grow closer together, and at 4,000 feet the forest is general everywhere. North of the Canadian Pacific Railway the dry country extends up the North Thomp-son and Fraser valleys, but within fifty miles on the former and one hundred miles on the latter, without much change in altitude, the country merges into the usual sub-arctic forest and the general flora takes on an eastern aspect.

The Coast Range, which extends from the Inter-national Boundary to Alaska, shuts out the humid winds of the Pacific and at the same time confines many western plants to a narrow strip along the coast. These with those found on Vancouver Island and the islands in the Gulf of Georgia and Queen Charlotte Islands, constitute a distinct flora in many respects.

Many genera peculiar to the west coast both to the

south and north and numerous species of other genera
fill the woods and open spaces with beautiful flowers,
and the spring months, April and May, are a season of
continual bloom. Liliaceous flowers are abundant, and
*Erythoniums, Trilliums, Alliums, Brodiæus, Fritil-
larias, Liliums, Camassias*, and others, are in great
profusion.

The space at my disposal will not permit me to go
into greater detail regarding the orders, genera and
species in the various districts, but my "Catalogue of
Canadian Plants" will give, to any one desirous of
obtaining them, details of the distribution of all
species known to occur in Canada.

PART II

HISTORY AND ADMINISTRATION OF CANADA.

PART II.

HISTORY AND ADMINISTRATION OF CANADA.

CHAPTER I.

Ethnology of the Aborigines.

BY ALEX. F. CHAMBERLAIN, PH.D., CLARK UNIVERSITY, WORCESTER, MASS., U.S.A.

IN an address before the American Association for the Advancement of Science at Nashville, Tennessee, in August, 1877, the late Sir Daniel Wilson, to whom the science of ethnology in Canada owes at once its academic recognition, and its popular extension, used these significant words: "The work before you for years to come must be the accumulation of evidence, the cautious sifting of it in all its bearings, and the ascertaining what its teachings really are."

Twenty years have elapsed since this plain statement of the task, and now a few, at least, of the perplexing problems have vanished, many extravagant theories have been consigned to the limbo of oblivion; ghost-tribes and ghost-languages have disappeared from the maps, while with increasing exactitude the boundaries of linguistic provinces and culture areas are being marked off and delimited, and the diffusion of physical stocks over the American continent investigated and determined with some approach to certainty.

The researches of faithful missionaries, painstaking travellers and investigators, illumined by the philosophic insight of Hale, Powell, and Brinton, have made

it clear that, for the present—until physical anthropology shall, perhaps, be able to afford us a better one—
the only safe classification of the Indian tribes of
America, is according to the languages which they
speak.

The American race in Canada (and Newfoundland)
can, therefore, be classified, as follows, each division
ranking as a separate and distinct family of speech :
I. Eskimo. II. Beothuks. III. Algonkins. IV. Iroquois. V. Sioux. VI. Athapascans. VII. Kootenays.
VIII. Salish. IX. Kwakiutl-Nootkas. X. Tsimshians. XI. Haida.

Eskimo (from a word in one of the eastern Algonkian tongues, signifying " raw flesh eater "). The
Norsemen came to Greenland in the tenth century, and
found there the Eskimo, this being the first known
contact of the American aborigines with visitors from
Europe. The extent of Norse voyages and explora-
tions has been unwarrantably exaggerated by certain
writers, for outside of Greenland the influence of these
early immigrants from the old world is imperceptible.
Within the last few years the researches of Rink and
Holm in Greenland, of Packard and Turner in Labrador, of Boas in Baffin Land, of Petitot in the delta of
the Mackenzie, of Dall, Petroff and Murdoch in Alaska,
have shed a flood of light upon the migrations, arts and
inventions, languages, customs and beliefs of this
roving maritime people, whose numerous settlements
have occupied the littoral of the Arctic from Greenland to Alaska and the Aleutian Islands, besides the
coast of Labrador down to the Straits of Belle Isle
(with occasional incursions into Newfoundland and
north-eastern Quebec), and the extreme north-east of
Siberia as far as the river Anadyr—the Eskimo
(Chukchi) of the last mentioned region being the only
aboriginal American intruders into Asia, so far as is
known. The most reasonable theory (Boas) makes the
early home of the various Eskimo tribes—whose

similarity in speech, and especially in mythological *fond*, is striking—to have been "in the lake region west of Hudson Bay," whence one branch wandered by way of Baffin's Bay to Greenland and Labrador, others northward and north-westward. The former southward range of the Eskimo is still a matter of dispute, but Abbot, Packard and others, suggest their descent from glacial man in America, who retired northward with the retreating ice-sheet. Boyd Dawkins (and later, Lubbock and Beddoe) seek to connect them, chiefly by analogies of weapons and art in bone, etc., with the men of the river-drift in France. As to the influence of the Eskimo upon the neighboring Indian tribes not much is known, and no linguistic affinities have as yet been discovered—Herzog's attempt to connect the Eskimo with the Yuma of Southern California, is a type of many useless efforts in this direction. Proofs of physical intermingling, of transference of inventions and of myths are, however, not lacking. Dr. F. Boas sees in the low indices of the Micmacs, and of ancient skulls from New England, evidence of an intermixture of Eskimo blood, while Prof. Putnam arrives at the same conclusion upon archæological grounds, and others, less rightly, from comparisons of myths. In certain archæological specimens from the Algonkian area in Ontario, Eskimo influence has been hinted (Boyle). According to Dr. Boas, the Eskimo have influenced the style of carving of the Tlingit, while the throwing-board of these Indians, like the harpoon of the tribes of the North-West Coast, is of Eskimo origin, though subjected to artistic modifications. Upon the Eskimo of Alaska, and those west of the Mackenzie, the Indians of the North-West have exercised considerable influence, as is seen in the use of masks, labrets, wooden hats, patlatches, singing houses, sweat baths, slavery, art of carving, etc. (Boas). It is even possible, Dr. Boas thinks, that the plan of the Eskimo snow-house can be traced back to some-

thing like the square house of the western tribes.
This influence seems to have extended, though but
slightly, to the Eastern Eskimo. The use of tobacco
and the pipe, the employment of nets in fishing, and
the use of the " bird bolas," are due, according to Murdoch, to borrowing from the natives of north-east
Siberia. The attempt to make of the Eskimo a race
entirely distinct from the other American aborigines,
so favored by certain writers, seems to have been no
more successful than the effort to derive them in recent
times from the Mongols of Siberia. The Eskimo are a
good-natured, imaginative, inventive, artistic people,
with a real talent for song and story, and a skill in
hunting and fishing, which makes their self-given
name—*Innuit*, "people"—not so unjustified as might
seem to be the case at first sight. The literature in
and about their language till the year 1887, is described
in the late Mr. J. C. Pilling's excellent Bibliography of
the Eskimo family of speech, while the histories of the
Moravian and Oblate missions and the archives of the
Russo-Greek church contain much useful general information, which can be compared with the results of the
more exact researches of later scientific investigators.

Beothuks (from a word said to mean " red," in the
language of these people). The Beothuks, or " Red
Indians," of Newfoundland, are now entirely extinct—
the last survivor is said to have died in 1829. All
that is known concerning them is to be found in the
recent writings of Dr. A. S. Gatschet and the Rev.
George Patterson. In arts and customs they seem, in
some respects, to have differed noticeably from their
neighbors, and, from the scanty records of their speech
which have been preserved, it is probable that we have
in them a distinct linguistic stock, with, however,
borrowings from Algonkian, and possibly also, Eskimo
sources. Dr. Patterson seems inclined to rank the
Beothuk language as Algonkian, but on very insuffi-

cient grounds. Dr. D. G. Brinton, detects a slight
resemblance in general morphology to the Eskimo.
The history of the extermination of this people by the
whites is a dark page in the annals of Newfoundland.

Algonkins (the name is a corruption of *agoomegwu*
which, in the allied Ojibwa and Nipissing dialects,
signifies "other side-of-the-water people," in reference,
probably, to the St. Lawrence). This wide-spread people, to whom belong the Naskapis and Scoffies of Labrador; the Micmacs of Nova Scotia and Prince Edward
Island; the Abnakis, Meliseets, etc., of New Brunswick;
the Passamaquoddies and kindred tribes of Maine; the
Nipmucks Narragansetts, Mohegans, and related peoples of New England and New York; the Lenâpé, on
the Delaware River; the Nanticokes and kindred
tribes in Maryland and Virginia; in the Ohio region,
the Weas and Piankashaws; on the Tennessee, the
Shawnees; in Illinois and the adjacent territory, the
Illinois, Kaskaskias, Kikapoos; around Lake Michigan,
the Menomonees, Pottawattomies, Sacs and Foxes;
about Lakes Ontario, Huron, and Superior, from the
Ottawa River to the Lake of the Woods, the Ottawas,
Nipissings, Ojibwas (many tribes), etc.; in the great
west of Canada, the Crees; on the Saskatchewan and
Missouri, the Blackfeet; on the Kansas and Arkansas,
respectively, the Arapahos and Cheyennes,—is one of
the most interesting and important of all American
Indian stocks. The dialects of the Algonkian speech
are very numerous; but the Cree is thought by many
to preserve the original mother-tongue best of all,
while the languages of the Blackfeet, in the extreme
west, and of the Micmacs, in the extreme east, depart
most from the parent stock. The members of this extensive family have wandered far and often, but eminent scholars, from a study of mythology, languages,
and archæology, seem inclined to place the early home of
the Algonkins somewhere "north of the St. Lawrence
and east of Lake Ontario" (Brinton). As may be seen

from the Bibliography of Pilling, which contains 2,245
titles of books, articles, and manuscripts, the literature
relating to this family of speech, from the times of the
Jesuit missionaries to the present, is quite voluminous,
dictionaries, translations of the Scriptures, transcripts
of legends and myths abounding, while several syllabar-
ies and special alphabets have been invented to aid in
transliteration. The Algonkins possess a rich mythology, of which the demi-god Nanibozhu (Glooskap with
the Micmacs, Wisketchak with the Crees, Napi with
the Blackfeet), is the central figure. The "medicine
societies" of the Ojibwa, studied by Hoffman, and the
sun-dance of the Blackfeet, investigated by Maclean
and others, are among the most interesting ceremonial
institutions. In their highest religious thought they
reached a stage of nature-interpretation akin to that of
the Vedas and the Edda (Brinton), and, in connection
with their shamanistic rites, had developed picture-
writing to a remarkably high degree—the Lenápé and
Ojibwa in particular. Agriculture (the raising of maize,
squashes, tobacco) was well under way when they first
came into contact with the whites, and among their chief
manufactures were pottery, mat-weaving, skin-dressing,
and the fashioning of arms and utensils of wood and
stone, while their river life and prairie travel made
them also traders of no mean sort. Deserving of closer
and more searching investigation are the totemic systems of the various Algonkian tribes and the history
of the Blackfoot confederacy. The typical Algonkins
are a physically well-developed people, more genial and
fanciful than is generally believed, and of great intellectual capacity and political abilities—Pocahontas,
King Philip, Pontiac, Tecumseh, Black Hawk, Crow-
foot, Poundmaker, were all Algonkins. The influence
of the Algonkins upon the language (more than a hundred words, *chipmunk, Tammany, sachem, mugwump,*
etc., have crept into the popular speech of the United
States and Canada from the various Algonkian dialects),

manners and customs, arts (maple-sugar making is of
Indian origin), agriculture (cultivation of the squash
and the use of fish-manure are due to the Indians of
New England), and amusements (lacrosse and the to-
boggan come from these Algonkins), of the European
immigrants and their descendants has been considerable, while intermarriages have been very frequent, especially with the French of Canada and the north-western States and the Scotch settlers in the Hudson's Bay
region, generally with improvement on both parents.
The half-breeds of Manitoba and the north-western
Provinces still form an important social and political
factor in the Dominion. The various tribes of the Algonkins have long been more or less under missionary
influence, and the settlement of the Mississagas at New
Credit, near Brantford, Ontario, is a marked example
of what has been done in civilizing the Red Man. Of
recent years the best special studies of Algonkian languages, mythology, institutions, and history, have been
made by Brinton (Delaware); Tooker, Gatschet, Trum-
bull (Algonkins of Virginia, Middle and New England
States); Leland and Rand (Micmac, etc.); Cuoq (Nipissing); Wilson, Horden, Hoffman (Ojibwa); Horden,
Watkins, Lacombe (Cree); Sims, Maclean, Grinnell
(Blackfoot).

   Iroquois (the derivation of the name is still uncertain). This family is perhaps the most remarkable, in
intellectual, as well as in physical development, of all
the Indian tribes north of Mexico. The recruits they
furnished to the Federal troops of the United States
during the war of the rebellion, " stood first on the list
for height, vigor and corporeal symmetry " (Brinton),
and Horatio Hale, a careful and thorough student of
these Indians, does not hesitate to say that " their
achievements, institutions, and language show them to
have been, in natural capacity and the higher elements
of character, not inferior to any race of men of whom
history preserves a record." The early home of the

Iroquois seems to have been "in the district between the lower St. Lawrence and Hudson Bay (Brinton)." and they were at no period of their history a maritime people; the shadow of the forest seems always to have been about them. To this stock belonged the Five Nations (Cayugas, Mohawks, Oneidas, Onondagas and Senecas) in New York and later (again) in Ontario; the Hurons (Wyandots) and the so-called "Neutrals," in Ontario; the Eries in the region south of the lake of that name; the Andastes and Susquehannocks on the lower portion of the Susquehanna River; the Tuscaroras in Virginia; and the Cherokees in Carolina and Tennessee River region. The identification of the Cherokee (to whom not a few of the mounds in their former territory are doubtless due) as a distant branch of the Iroquoian family of speech is owing to Horatio Hale, whose deep studies of Iroquois sociology have resulted in demonstrating the great political and administrative capacity of these Indians, qualities in which they far surpass all the uncivilized races of the New World. Their general council with its recognition of the conservative-aristocratic, and the elective-democratic elements, their balancing of federal privileges and local rights, their adoption of descent in the female line and woman suffrage, and their high respect for women,—the peace-makers for the people,—were enough to win for this gifted people the admiration of every investigator. But more than this, some four hundred and fifty years ago, an Onondaga chief, Hiawatha,—now happily rescued from the realm of myth and legend,—a wise and prudent statesman, formed a remarkable plan,—styled by Dr. Brinton "one of the most far-sighted, and in its aim beneficent, which any statesman has ever designed for man,"—for a federal union, with tribal autonomy of his nation with the Mohawks, Oneidas, Senecas, and Cayugas. The object of this very successful league, to which were admitted afterwards the Tuscaroras, and even portions of other

tribes, was to abolish war and institute peace for ever.
It is a credit to American Indian humanity that this
federation, born of a people too commonly known as
the fierce and cruel Iroquois (as a penalty for the savage
deeds of some of them towards the whites and their
own kindred, the Hurons), lasted for more than two
centuries, to crumble away at the coming of the whites.
Some of the Iroquois tribes had women chiefs or
" Queens," and among the Wyandots, whose system of
government has been sketched by Powell, we learn
that "the council of each gens was composed exclusively
of women. They alone elected the chief of the gens
who represented its interests in the council of the tribe "
(Brinton). The Iroquois (Cherokees especially) showed
considerable capacity as agriculturalists, and their
house architecture (that of the communal houses in particular) was quite beyond the ordinary. The " Book
of Rites," embodying the national songs of the Iroquois
has been edited by Mr. Hale, and Mr. James Mooney
has collected and published the " Sacred Formulas " of
the Cherokee Shamans. In the mythology of the Iroquois, often of a rather philosophical, or even metaphy-
sical cast, animal tales are quite numerous, but the
typical legend is that of the twin brothers,—" Good
Mind " and " Bad Mind," the old story of the battle of
light and darkness. Among those who, in the last few
years have made special studies in Iroquoian myths
and folklore, may be mentioned Mrs. Erminnie A.
Smith, Horatio Hale, Rev. W. M. Beauchamp, James
Mooney, J. N. B. Hewitt. The invention of the Cherokee syllabary (now in use in the Indian Territory
among the Cherokees settled there) by George Guess,
or Sequoyah, a half-breed Cherokee, about the year
1821, is a memorable event in Indian history. Besides
Hiawatha, the Iroquios stock has produced men like
Brant, Red Jacket, and Oronhyatekha, the last a man
of notable organizing and executive capacity, at present
the head of a great friendly society. The barbarism of

s

the Iroquois in their wars with the Hurons, and their
cruel torture of the Jesuit missionaries, are dark blots
on their escutcheon, but at Caughnawaga, St. Regis, at
Oka, Thyendinaga, on the Thames, and on the Grand
River, their descendants live to-day industrious farmers and expert mechanics, the equals (sometimes the
superiors) of the whites around them.   The Indians of
the Six Nations Reserve near Brantford, Ontario, chiefly
the descendants of those Iroquois, who crossed to Canada from New York, after the conclusion of the war
of 1776, held, in July 1896, a highly successful exhibition, under the auspices of their Agricultural Society
founded in 1867, and in 1885 celebrated with great
éclat the centenary of their settlement on the Grand
River. The literature in and about the Iroquoian family
of speech, from the Huron vocabulary of Cartier (in
1545) down till 1888, is described in Pilling's Bibliography.

    Sioux (from Nadouessioux, a French form of an
Algonkian name of this people). The Dakotan or
Siouan family of speech includes the dialects of the
Sioux or Dakotas proper, on the Upper Mississippi and
Missouri, and (a few) in Manitoba and the Canadian
North-West; the Assiniboins (Stonies), on the Saskatchewan and Assiniboin Rivers; the Mandans and Pon-
kas, on the Middle Missouri; the Crows and Minetaris,
on the Yellowstone; the Ottoes, on the Platte; the
Arkansas, Quapas and Osages, on the Rivers Arkansas
and Osage; the Omahas, on the Elkhorn; the Kansas,
on the Kansas; the Iowas and Yanktons, on the Iowa;
the Winnebagos, on the western shore of Lake Michigan; the Biloxi, in Louisiana; the Catawba, etc., in
the Carolinas; the Tutelos, in Virginia.   The Canadian
representatives of this family are comparatively recent
intruders.   An investigation of the speech of a Tutelo
resident on the Indian Reserve near Brantford, Ontario, by Mr. Horatio Hale, led him to see in it a distant relative of the Sioux, and later on competent stu-

dents determined the Catawba to belong also to the
same stock, likewise the Biloxi (Dorsey). The reasonable supposition that the early home of the whole
Siouan family was in the region of the Carolinas, a conclusion based upon the study of these languages, became
a certainty through the thorough examination of their
mythology by the late Rev. J. Owen Dorsey, the best
authority on all matters relating to these people. It
will be seen from this how far the Canadian Sioux
have wandered from their primitive habitat. The
Bibliography of Mr. Pilling, published in 1887, reveals
a wealth of literature in and about the languages of
the Siouan stock. The typical Sioux Indians are
strong and well-built and their fame in war is well-
known ; the majority were nomadic hunters and fish-
ers of no mean skill—the Mandans, however, and some
others, were more sedentary and devoted to agriculture. Their gentes system, government and laws have
been described at length by Mr. Dorsey and Miss Alice
Fletcher, who have also investigated deeply their ceremonials, customs and beliefs, secret societies, dress,
ornaments, arts, inventions. With these Indians, picture-writing had reached a highly developed stage,
and their names for objects in nature and for the attributes of human personality often evince naive philosophic keenness of thought (Fletcher). In their mythology, animal tales and stories of giant monsters
abound—the water-spirit, the thunder-bird, the rabbit, and Ictinike, the incarnation of maliciousness.
The most recent writers who have made special studies in the ethnology of these people are Rev. J. Owen
Dorsey, Rev. S. R. Riggs, Rev. J. W. Cook, Miss Alice
Fletcher, Horatio Hale, James Mooney,—an essay by
the last mentioned, "Siouan Tribes of the East," containing the results of very recent (1894) investigations.

Athapascans (the meaning of the term is uncertain
—derived from Athabasca, the lake name; they are
also known as Déné, which signifies "men"). The

geographical distribution of this stock is very remarkable. The family includes the various tribes known
as Loucheux in Alaska and the adjacent Canadian territory; the Hare Indians (about the Mackenzie, Anderson and McFarlane Rivers) ; the Bad People (at Old
Fort Halkett); the Slaves (west of Great Slave Lake);
the Yellow Knives (north-east of Great Slave Lake);
the Dog-Ribs (between Great Slave and Great Bear
Lakes); the Cariboo Eaters (east of Lake Athabasca) ;
the Chippewyans (about Lake Athabasca); the Tsé'
kéhne (on both sides of the Rockies)—to which
belong also the Beavers (south of Peace River) ; and
the Sarcees (east of the Rockies about 51° n. Lat.) ;
the Nahrane (of the Stikeen River) ; the various tribes
known as Carriers (about the region of Stuart's Lake) ;
the Tsilkoh'tin (on the Chilcotin River)—these form
the northern Déné group (Morice) ; the southern Déné
are : the Upper Umpquas, the Micíkqwûtmétûnnĕ of
the Upper Coquille River, the Chasta Costa, north of
Rogue River, the Chetcho, south of Rogue River,
and some other settlements in this portion of Oregon ; the Kwalhioqua, in north-western Washington
(Powell) ; the Qáamóténe (Smith River Indians), and
a few tribes south of Smith River, in north-western California ; the Wailakki, along the western slope of the
Shasta mountains, in the same state ; the Hupá, on
Trinity River, California ; the Navaho, in northern
New Mexico and Arizona, with offshoots in Colorado
and Utah ; the Apache, in New Mexico (with offshoots
into Mexico), Arizona, Colorado, and (recently) Okla-
homa ; the Lipan (formerly roving from the Red River
to the Rio Grande), now in New Mexico and Mexico. The contrast between some of the very rude
Athapascan tribes of the north and the artistic and
progressive Navaho is well brought out by Horatio
Hale in his remarkable essay on "Language as a Test
of Mental Capacity." Among no other people, perhaps,
can be seen so well exhibited the results produced by

differing environment. Equally striking is the contrast between the warlike Apaches (or the Hupâ—the Romans of California, Mr. Powers calls them) and some of the tribes of the great Barrens of north-western Canada. Perhaps no portion of Canadian ethnology is so interesting, as the study of this numerous family, whose many dialects, extensive range, and varying stages of barbarism and culture, afford a wide field for careful and serious investigation. Among most of the Athapascan tribes, arts, institutions, religions and languages, seem equally primitive, but the Navaho and Apache gentile system, studied recently by Bourke and Matthews, seems more complicated, and the legends connected with the origin of the gentes are numerous and detailed. The Navaho mythology—with its "songs of sequence," "Mountain Chant," and many sacred narratives, collected and published by Dr. Washington Matthews, together with their known skill in weaving, picture-writing, agriculture, etc., give abundant evidence of the capacity of the Athapascans under favorable circumstances and in less severe environments. The early home of the family seems to have been somewhere in north-western Canada. The best recent studies of the Athapascans have been made by Petitot, Bompas, Kirkby, Legoff, McDonald (Canadian North-West); Morice (British Columbian tribes); Dorsey (Athapascans of Oregon); Powers (Californian tribes); Bourke (Apache); Gatschet (Lipan): Matthews (Navaho). The extensive literature (the result of missionary labors, chiefly in the north) in and about these Indians, may be found titled and described in Mr. Pilling's Bibliography, published in 1892. In the Report of the British Association for 1895, Dr. F. Boas describes a new and interesting Athapascan people, the Ts'ets'áut, on the Portland Canal, in British Columbia. The headform of the Athapascans, in general, is brachycephalic, and the influence of their intermixture with other tribes can be traced among the Tlingit of Alaska, the

Bilqula of British Columbia, and certain peoples even
as far south as Northern California. Points of mytho-
logic contact are also observable. The investigations
of Rev. A. G. Morice, among the Déné Indians of the
Stuart's Lake region, are resulting in the accumulation
of data that will help much in the solution of the problem
of the influence of the Athapascans upon the people of
the North-West, while Boas' studies of the Salish and
Kwakiutl-Nootka, afford a sure basis for comparison.
A careful anthropometric investigation of the Athapascan tribes of the Canadian North-West is very desirable,
and the Beavers and the Sarcees, especially, are worthy
of further study.

Kootenays (the meaning of the word Kitonaqa,
applied by these Indians to themselves, is uncertain).
This people, who inhabit the south-eastern part of
British Columbia, northern Idaho, and north-western
Montana,—the region watered by the Upper Columbia
and the Upper Kootenay Rivers—speak a language in
two slightly differing dialects (the Upper Kootenay and
the Lower Kootenay), which ranks as a separate linguistic stock. Their earlier home, as stated in their
traditions, was to the east of the Rocky Mountains,
whence they were probably driven by the Blackfeet,
with whom they were a long time at war. With the
Blackfeet and the neighboring Salish—a little colony of
Shushwaps resides on the Columbia Lakes, quite within
the Kootenay Territory—they have intermarried somewhat, and probably borrowed something in customs and
myths. From the time of the missionary De Smet,
down to the present day, all who have really understood these Indians have taken a very favorable view
of their character, and the industrial school at St.
Eugène, in the Upper Kootenay region, seems to thrive
under the careful administration of the devoted
missionaries and nuns of the Catholic church, whose
influence generally, upon these Indians has been productive of lasting good—a drunken Kootenay being a

rare thing indeed. The government of the Kootenays
is very democratic in character, and there appears to be
a complete lack of totem-systems and secret-societies.
Although artistically inclined, as their weapons and
utensils show, picture-writing seems not to have
appealed strongly to them. They have a rich and
interesting mythology, the chief feature of which consists in animal tales, the coyote, chicken-hawk, grizzly
bear, playing the principal rôles. In recent years the
Kootenays have been visited and studied by Dr. Franz
Boas and the present writer. An account of their
ethnology, mythology, physical characteristics and
language, by the latter, appears in the Report of the
British Association for 1892.

Salish (the etymology of the name is uncertain).
This family embraces a large number of tribes resident
in the North Pacific coast region, between the fifty-
third parallel, and fifty miles to the south of the
Columbia, where a Salish colony, the Tilamuks dwelt
in Oregon. The principal members of the stock are, in
British Columbia the Bilqula (Bella Coola), on Dean
Inlet and Bentinck Arm; the Coast Salish (the Catloltq,
Siciatl, Péntlatc, Skqōmic, Kaiutcin, Lkéngen); the
Ntlakyápamuq, in the interior, between Spuzzum and
Ashcroft; Stlatliumq, in the region of Douglas and
Lillooet Lakes; the Squápamuq (Shushwap), in the
region of Kamloops and Shushwap Lakes; the Okinā-
kēn, around the Okanagan and Arrow Lakes; in the
State of Washington, the Niskwalli, Twana, Qtlummi
(Lummi), Tscēlis (Chehalis), Tlallam, Samish, Kwinautl,
and a few other small tribes, besides the so-called
Flatheads, Kalispelm, Colvilles, Pend d'Oreilles, stretch-
ing over into northern Idaho and north-western
Montana; in Oregon, the Tilamuks, on the coast about
fifty miles south of the Columbia. Since Gibbs, the
chief investigator of the Salish peoples (of British
Columbia in particular) has been Dr. F. Boas, the
results of whose extensive anthropometric, ethnologic

and linguistic researches have appeared in the various Reports of the British Association 1889-1896; the Shushwap have also been studied by Dr. G. M. Dawson. Among the missionaries and others who, of recent years, have written in and about the languages of the Salish family, may be mentioned Eells (tribes of Puget Sound), Giorda and Canestrelli, the successors of Mengarini (Kalispelm), Le Jeune (Shushwap), Tolmie and Dawson, Boas (various tribes of the mainland of British Columbia and Vancouver Island. The Bilqula (comparatively recent immigrants), studied in detail by Boas, are interesting as a Salish tribe entirely surrounded by alien stocks, and like the outlying Tilamuk of Oregon, deserved special attention. The literature of the Salish languages till 1893, is recorded in Mr. Pilling's Bibliography. In many respects the Salish are the most important of all the Indian tribes of British Columbia, the diversity of dialects, divergence of customs and ceremonies, unlikeness of culture, indicating much migration and influence of other stocks,—the Bilqula (particularly in mythology) have borrowed much from the Kwakiutl. Some relationship between the Salishan and Kwakiutl —Nootkan linguistic stocks has been suggested, but proof is not now in evidence. The mythology of the British Columbian tribes of the coast, forms the subject of a volume published in 1895, by Dr. Franz Boas—a collection of myths and legends, which must long remain our chief source of information on the subject. Equally valuable are the anthropometric data contained in the Report of the British Association for 1895 —by the same writer—and his studies of the social system of the Coast and Inland Salish.

Kwakiutl-Nootkas. The language of the Kwakiutl Indians embraces three dialects, the Qāislá (north of Grenville Channel), the Hēiltsuk (from Grenville Channel to Rivers Inlet, both on the north-eastern coast of British Columbia, and the Kwakiutl proper, spoken further south in the same region, and on the northern portion of Vancouver Island. The language of the Nootkas comprises the dialects of some twenty-two tribes or settlements, chiefly in the west coast region of Vancouver Island (Barclay, Clayoquaht, and Nootka Sounds), and including the Makah and related tribes in the State of Washington (Cape Flattery, San Juan Harbor, Nitinat Sound). In recent years (Sproat's account of the Nootka, which appeared in 1868, is still, in many respects, admirable, as is also Swan's treatise on the Makah, published in 1868), the Kwakiutl language and people have been studied by Rev. A. J. Hall (whose grammar appeared in 1889) and Boas, the Nootkas by Boas, the latter also furnishing in the Reports of the British Association for 1889, 1890 and 1896, the best account of the ethnography and ethnology of both peoples. In the Report for 1890, Dr. Boas brought forward evidence to show that the Kwakiutl and Nootka belonged to one linguistic stock, a conclusion now generally accepted. The same authority considers as of some weight the fact that the Kwakiutl, Salish and Chemakum (of northern Washington) languages, all have pronominal gender (a rare phenomenon in American aboriginal speech), which, together with a few other structural peculiarities, suggests some remote relationship between these stocks, though no such evidence can be gleaned from the vocabularies. To Dr. Boas we owe our knowledge of the intricate social organization, and numerous secret societies of the Kwakiutl, which, like the *potlaches* of the Nootkas and their neighbors, the dances of many British Columbia tribes, are fertile fields for further investigation. The same investigator has made a special study of the houses of the Kwakiutl—a topic, the careful consideration of which is important for the understanding of the culture-borrowings of the peoples of the north-west coast and northern Canada. Between the Kwakiutl and the

Bilqula, considerable borrowing of words has taken place, and *vice versâ.* The religious ceremonials of the Kwakiutl have made their influence felt even in Alaska, while, by reason of marriage-customs, "a large number of the traditions of the neighboring tribes have been incorporated into their mythology" (Boas). The literature in and about the Kwakiutl-Nootka linguistic stock, is described in Pilling's Bibliography of the Wakashan family, so the authorities of the Bureau of Ethnology at Washington style this people. Worthy of special study are the social institutions of the Hëiltsuk and the Kwakiutl, which differ radically.

Tsimshian (the name signifies "on the Ksian'n," *i.e.,* Skeena River). The Tsimshian linguistic stock is composed of two dialects, the Nasqa and the Tsimshian proper, spoken in the region of the Nass and the Skeena Rivers in north-eastern British Columbia. The religion of these Indians is "a pure worship of heaven" (Boas), but their secret societies and shamanistic rites show Kwakiutl influence, while some of the latter have been borrowed by the Haida from the Tsimshian. The best account of Tsimshian ethnology is given by Dr. Franz Boas, in the Reports of the British Association for 1889, 1895, 1896. A study of the Tsimshian language was published by Count von der Schulenburg in 1894. The Tsimshian, like the Haida and the Tlinkit, is noteworthy as possessing an *r*-sound, and it is also marked off from the more southern languages by the almost exclusive use of prefixes (suffixes being very rare in occurrence),—its principles of composition differing radically from those in use in the Kwakiutl-Nootka, etc. It is among the Tsimshian Indians that Mr. Duncan has labored so long and so successfully. The story of his mission at Metlakahtla ("The Story of Metlakahtla," by H. S. Welcome, London, 1887), is one of the most remarkable records of colonizing and proselytizing zeal in existence. In 1887, difficulties arose between the civil and ecclesiastical authorities and Mr. Duncan, the result of which was that many of the Indians followed their leader over the border into the United States Territory of Alaska, where, at Port Chester, on Annette Island, a new settlement was founded.

Haida (the name really signifies "people"). The dialects of the Haida linguistic stock are spoken on the Queen Charlotte Islands, and part of Prince of Wales Archipelago, Alaska, by five tribes or settlements (according to the Rev. C. Harrison, there were formerly thirty-nine), at Masset, Skidegate, Gold Harbor, Houkan and Cassan—the last two being in American territory. A good account of the Haidas of Queen Charlotte Island was published by Dr. G. M. Dawson, and a grammar of the language, by Rev. C. Harrison, appeared in 1895. Contributions to the ethnology of this group have also been made by Dr. F. Boas, in various Reports to the British Association. Some relationship between the Haida and the Tlingit languages has been suspected. From the Tsimshian the Haida have borrowed several customs and dances. Among all the natives of the North-West coast the Haida are most distinguished for their totemic carvings and conventionalized ornamental art. The development and distribution of these forms from the Tlinkit of Alaska to the Makah of Cape Flattery, is a subject of the deepest interest to the ethnologist. A general account of the art of the northwest coast will be found in Lieut. Niblack's "The Coast Indians of Southern Alaska and Northern British Columbia," published by the U. S. National Museum in 1890, and several more recent contributions are due to Dr. Boas. The general ethnology of the Tlinkit, Haida and Tsimshian is treated of by Niblack. The literature in and about the Tlinkit, Haida and Tsimshian stocks is detailed in Pilling's "Proof Sheets of a Bibliography of the Languages of the North American

Indians," published in 1885—the special bibliographies not appearing since the death of Mr. Pilling, in 1895.

Chinook Jargon. Not to be classified with any of the linguistic groups of the Dominion, but more interesting in some respects, than any of them, is the "Chinook Jargon," or Trade Language of the Oregon—Columbia region. This *lingua franca*, the rise of which dates from the early years of the present century, owes its birth to the necessities of commerce between the whites and Indians, but has also been adopted as a means of communication between Indians of different speech to such an extent that it is said there are " few tribes between the 42nd and 57th parallels of latitude" (Powell), among whom some knowledge of it is not to be found, and its influence is to be traced even beyond the Rockies among the Algonkian Blackfeet and Crees. This jargon consists of words from several Indian dialects of the region (a few Algonkian terms, introduced by the French-Canadian half-breeds, also occur), English, and Canadian-French, many of them very much disguised and changed in form and meaning—grammar is almost *nil*, accent and intonation counting for a great deal. One of the Indian languages which has contributed to the jargon is the Chinook, formerly spoken on the Columbia from the Cascades to the Pacific. This linguistic stock has been best studied by Dr. Boas, the jargon by Eells, Le Jeune, Horatio Hale, and Boas, a dictionary by the third eminent authority having appeared in 1892. The literature in and about the Chinook jargon and the Chinook language will be found titled and described in Pilling's Bibliography (1893). On more than *vivâ voce* grounds, the Chinook jargon must be counted among the living tongues of Canada, for Father Le Jeune, the excellent missionary among the Shushwaps, has published at Kamloops, since May, 1891, the " Kamloops Wawa," a weekly in the jargon, aided by the Duployan system of shorthand, introduced among these Indians by the editor.

General Considerations and Problems. It is noteworthy that among these northern peoples—Tlingit,
Tsimshian, Haida—the custom of compressing the head,
which has given a name to the Flatheads (Salish) of
Montana, seems never to have prevailed. Besides
members of the eleven stocks above considered, stray
Indians from over the border have passed into Canada
occasionally—a few Shoshonees, *e.g.*, from Montana.
In the Report of the British Association for 1895, Dr.
F. Boas discusses in detail the physical characteristics
of the Indians of the north Pacific Coast region, recognizing four types in British Columbia, as follows:
Nass River Indians, Kwakiutl, Harrison Lake Indians,
and Interior Salish, Okanagan, Flathead, Shushwap.
The relation of these to the types of the region from
Alaska to Mexico on both sides of the Rockies is not
yet clear. The Athapascans probably belong to the
North Pacific group, while there are certain reasons
for grouping together the Algonkin, Iroquois and doli-
chocephalic Eskimo. The inadequacy of the census
of some of the northern and western tribes leave
somewhat doubtful the question whether the Indians
of Canada have not considerably decreased of recent
years, as many good authorities believe. A fair
estimate of the aborigines belonging to the different
stocks seems to be : Eskimo, 6,000 ; Algonkins, 60,000 ;
Iroquois, 9,000 ; Sioux, 2,000 : Athapascans, 17,000 ;
Kootenays, 700 ; Salish, 12,000 ; Kwakiutl-Nootkas,
5,000 ; Tsimshians, 4,000 ; Haida, 2,500, a total of 118,-
200. Dr. D. G. Brinton (in a recent communication to the
present writer) instances as perhaps the most important
topics in the anthropology of Canada : (1) The ethno-
graphic relations of the north-west coast tribes ; (2)
The thorough investigation of the emblematic and
other mounds of Ontario ; (3) The investigation of the
shell-mounds, village-sites, and aboriginal cemeteries of
the eastern coast (Quebec, Labrador, New Brunswick,
Nova Scotia), with special reference to (a) antiquity

(palæolithic man), (b) areas of culture (Eskimo, Algonkian), (c) relics of Northmen. The first of these topics has been touched upon quite recently by Boas and Mason, who incline, the first from mythological, the second from cultural considerations, to see Asiatic influence in this part of America. No case for Asiatic influence south of Alaska is yet made out, and Winkler, in his Ural Altaic studies, has recently shown how weak the attempts are to connect the Eskimo with the languages of Northern Asia. The second topic has been treated in the Archæological Report of Mr. David Boyle (whose investigations form the solid basis of Canadian archæological studies), for the year 1896-7, and now that the "glamor of the vanished mound-builders" has passed away, we may hope for real advancement along this line. The third topic has become increasingly important, but little real work has been done in the lines suggested, though the investigations of Piers and Patterson have been of general value. The call for unprejudiced students of Canadian ethnology should not be heard in vain, and the next few years should see many valuable contributions to the solution of these interesting and important problems in the history of American man.

CHAPTER I.—APPENDIX A.

Note on the Ethnical Affinities of the Present Inhabitants of Canada.

By Professor James Mavor.

THE earliest immigrants from Europe into the region now forming the Dominion of Canada were French. The coming of these is described by Mr. Benjamin Sulte, in the succeeding note. The French pioneers not only occupied portions of what is now Quebec; but they penetrated very early into the wilderness of the west. Their descendants have formed the bulk of the population of the lower St. Lawrence, have contributed to the pioneer settlement of newer provinces in the west, and have migrated in large numbers to the United States. The rapid increase of the French population in Lower Canada, the consequent extreme subdivision of the land, and the scanty industrial employment in their native province, have been the principal causes of this migration. The French and other pioneers in the west were rather trappers and hunters than settlers; and their mode of life led them to mingle readily with the Indian tribes among whom they found themselves. The result has been the growth of a half-breed race known as *Metis*. These form a proportion of the population of the North-West Territories and Manitoba which diminishes with growth of the white population by immigration and natural increase.

While the bulk of the population of the Maritime

Provinces, Ontario, Manitoba, the North-West Territories and British Columbia, are of British origin, yet the population comprises representatives of numerous races, and the recent colonization of Manitoba and the Territories has resulted in a series of very widely divergent racial groups.

The English form the bulk of the population of Ontario; Yorkshire and Devonshire in particular having sent large numbers of emigrants to the central part of the province. British Columbia is also mainly peopled by persons of English descent, while in the North-West Territories most of the ranchmen are English. The Scotch in Canada have exhibited their customary clannishness by settling to a large extent in definite groups. Nova Scotia is mainly peopled by Scots, and in Ontario, the counties of Bruce in the Huron Peninsula, and the counties of Grey and Glengarry are almost exclusively Scotch settlements. While the Scotch emigrant seems to prefer a rural life, the Irish emigrant seems to prefer the large towns. Although the total number of population of Irish extraction in Ontario is rather less than that of the Scotch population in the same province the number of Irish in Ottawa and Toronto is twice as great as the number of Scots in these cities. The Irish also form a large proportion of the population of New Brunswick.

In the Maritime Provinces and in Quebec the people of origins other than French and British form a small fraction of the population; but in Ontario there is a considerable German population, settled chiefly in the counties of Waterloo and North and South Renfrew, and in the city of Toronto. In the city of Toronto also there are over 600 Russians, for the most part Russian and Polish Jews, and about the same number of Italians. A few families of Spanish Jews have settled in Ontario. The negro population of Canada is not large. There are, however, several groups in the Niagara peninsula.

In Manitoba there is a large Scotch colony at Deloraine; there are German colonies at Morden and other places in Southern Manitoba ; Russian and Gallician emigrants have been settling on the line of the Manitoba and North-Western Railway. A prosperous colony of Mennonites* is settled at Gretna, and there are large colonies of Icelanders at Glenboro' and at Icelandic River and Gimli, on the western shore of Lake Winnipeg. The Mennonites have recently been mingling with the population round them, and their absorption is probably only a question of another generation or two at most. The Icelanders have not as yet exhibited this tendency to any manifest extent.

The most heterogeneous population of any portion of Canada is to be found in the North-West Territories. Since the opening of the Canadian Pacific Railway in 1885, the Territories have been receiving the bulk of the stream of immigration from Europe, and have also been partly peopled by migration from North Dakota and from other States to the south of the international boundary. In the town of Lethbridge, in southern Alberta, where there are coal mines, there is an exceedingly mixed population for a town of about 2,000 inhabitants. There one may find, in addition to persons of British origin, Belgians, Italians, Norwegians, Swedes, Icelanders, French, Austrians, Germans, Dutch, Danes, Hungarians, Russians, Negroes, Chinamen, and Half-breed Indians. A colony of Russian Jews was established by Baron Hirsch at Hirsch in Assiniboïa ; but the invincible desire to peddle rather than to labour has drawn many

---

*Originally Dutch, the Mennonites migrated to Prussia in the sixteenth century, and later to Russia (MacKenzie Wallace, Russia, p. 372). The Manitoba group, which began to arrive in 1874, came chiefly from the Government of Ekaterinoslav, in South Russia. On the history and religious belief of the Mennonites see Kurzgefatszte Geschichte und Glaubenslehre der Altevangelischen Taufgesinnen oder Mennoniten, by C. H. A. van der Smissen. St. Clair, Ill. [U.S.A.], 1896.

9

of the immigrants from the farm. The principal foreign colonies are situated in northern Alberta, on
either side of the branch line of the Canadian Pacific
Railway. There are six German colonies in this
region, the total number of souls (July, 1896), being
1,400. At Edna in the same district there is a colony
of Austrians and South Russians numbering 270.
Near Wetaskiwin there are 760 Scandinavians, and
in the same region, about 100 Belgians. There also is
a colony of Moravians and a small colony of French.
The system of settlement and the choice of the colonists themselves have so far conspired to isolate these
colonies from one another and from the rest of the
population. Sooner or later they will, no doubt, amal-
gamate, but at present they have shown little disposition to do so.

In British Columbia there are about 9,000 Chinese
and a much smaller number of Japanese. These immigrants do not mingle with the European population.
In the mining districts on the frontier there is at present an extensive immigration from the United
States. Although some of the immigrants are of
European birth, most of them have been born in
America. At Cardston, in Southern Alberta, there is
a prosperous settlement of Mormons from Utah.

CHAPTER I. APPENDIX B.

Note on the Settlement of New France.

By Benjamin Sulte, F.R.S.C., Ottawa, Author of "Histoire des Canadiens-Français."

IT is not possible at present to indicate precisely the dates of the arrival of the French settlers in Canada, because no registers nor lists of immigrants were ever kept. The settlers came as a rule individually or in little groups of three or four families related to each other, as immigrants from various countries do at the present day.

From an examination of family and other archives extending now over thirty years, I make the following deductions :—

Perche, Normandy, Beauce, Picardy, and Anjou, contributed about 200 families from 1633 to 1663; by natural growth these families contributed in 1663, a population of about 2,200 souls.

In 1662-3 there came about 100 men from Perche and 150 from Poitou, Rochelle and Gascogne—with a small number of women.

The trade of Canada, then altogether in fur and fish, left Dieppe and Rouen in 1662-3, for Rochelle, and this accounts for the immigration from the latter port and from Poitou.

Britany never traded with Canada and never sent any settlers to the colony, although there is a widespread belief to the contrary.

From 1667 till 1672, a committee was active in Paris, Rouen, Rochelle and in Quebec to recruit men, women and young girls for Canada. This committee succeeded in effecting the immigration into Canada of about 4,000 souls.　Half of the girls were country girls from Normandy, and the other half were well educated women who did not go into the rural districts but remained in Quebec, Three Rivers and Montreal.

Since these people were brought to Canada by the organized efforts of a committee, we might expect to find some detailed record of their arrival; but as yet no such record is known to exist.　We are merely told by contemporary writers, how many arrived at such and such a date, and where most of them came from; but that is all.　The church registers, notarial deeds, and several classes of public documents, show abundantly the places of origin of these immigrants, but not the date of arrival.　The Provinces above mentioned have furnished that last immigration, to which Paris also contributed a good share.　The remainder of France (all the South and the East) had no connection with Canada.

The regiment of Carignan came to Canada in 1665 and left in 1669 with the exception of one company which eventually was disbanded here.

In 1673, the King stopped all immigration; and this was the end of French colonization.　The settlers remained on the soil happily, and in 1680, the whole population of Canada was 9,700 souls.　Double this figure every thirty years and we have approximately the present population of the Province of Quebec and that of the groups established now in the United States.

In regard to the troops and their disbandment in this country, much misunderstanding exists.　The fact is, that from 1673 till 1753, the garrison of Canada consisted as a rule of about 300 soldiers in all, under a captain of infantry and sometimes under a major, the latter being usually no longer young.

During the years between 1684 till 1713, on account of the war, an addition of six companies was sent to

Canada. The body of troops in the colony was officially styled "Détachement de la Marine," because they were paid and equipped by the Department of Marine and Colonies; but none of them were in the Marine. When some of these soldiers made up their minds to settle in the country, they received their discharge from military service and became "habitants."

Acadia was peopled without any kind of organization between 1636 and 1670, or thereabouts. No one has yet satisfactorily demonstrated where the French of Acadia came from, though their dialect seems to indicate their place of origin as the neighbourhood of the mouth of the Loire.

CHAPTER II.

A Sketch of the History of Canada.*

By GEORGE M. WRONG, M.A., PROFESSOR OF HISTORY IN THE UNIVERSITY OF TORONTO.

BEFORE the beginning of the seventeenth century no European State had made successful colonizing efforts in North America. Repeated attempts had been made. For the first of these, we must go back to the eleventh century. It is scarcely doubtful that, by way of Iceland and Greenland, Leif Ericson and other Northmen reached the shores of Canada, nearly nine hundred years ago; soon, however, their settlement disappeared, and later voyages had no connection with these remote and futile efforts.†

Circ. 1000.

Nearly five hundred years after the Northmen, John Cabot, stirred by the news of Columbus's success in southern regions, sailed out into the Northern seas under English auspices. Reaching land

1497.

---

* Mr. Francis Parkman made the history of French rule in Canada the labour of his life. His works upon this subject include eleven volumes, published between 1851 and 1892 (Boston, Little, Brown & Co.). His style is vivid, and the volumes, though sometimes defective in critical quality, are by far the most valuable contributions to Canadian history which have yet appeared. Dr. William Kingsford is engaged upon a very full history of Canada, of which eight volumes have been published, bringing the work down to 1815 (Toronto, Rowsell & Hutchison, 1887-1895). Dr. J. G. Bourinot, has recently published a short history of Canada in the "Story of the Nations" Series (New York, Putnam, 1896).

† A critical account of the voyages of the Northmen will be found in Fiske, *Discovery of America*, Vol. I., Chap. II. (Boston, 1893). *The Finding of Wineland the Good*, Ed. by A. M. Reeves (Clarendon Press, 1890), contains a photographic reproduction of "The Saga of Eric the Red," together with an English translation.

—we know not exactly when or where—he raised the English flag and claimed the shores that he touched for Henry VII. England did nothing effective for a hundred years, yet Cabot's exploit gave her a shadowy claim to North America, to be realized in the final struggle with France, which ended in 1763.*

Like England, France established no colonies in America in the sixteenth century. It was Portuguese and not French ships that hovered first about the entrance to the Gulf of St. Lawrence. In 1524, Francis I., in defiance of a Papal Bull dividing the New World

1524. between Spain and Portugal, sent out an expedition under a Florentine, Verazzano, but the only result was a meagre description of the

1534. Eastern Coast of North America, along which Verazzano sailed. Ten years later the attempt was repeated. Jacques Cartier, a seaman of St. Malo, passed through the northern entrance to the Gulf of St. Lawrence and sailed up the great river as far as

1535-6. what is now Quebec. He returned to France, but came out again in the following year, pressed further up the river, and after many days, reached a thriving Indian village, where now is Montreal, and where, too, angry rapids prevented further progress by boat. He spent a wretched winter on the banks of the St. Charles near Quebec, and in the spring carried the remnant of his company to France.

1541-3. In 1541 the effort at colonization was again repeated. It failed, and the curtain falls for sixty years on French rule in Canada.†

---

*Mr. S. E. Dawson in his *Voyages of the Cabots* (Pres. & Trans. Royal Soc. Canada, 1894), refers to all the principal works upon the subject. These are numerous, but few of them are in literary form.

†F. Parkman in *Pioneers of France in the New World* (Boston, 1865), gives an interesting narrative of French attempts at colonization in Canada, Brazil and Florida, in the sixteenth century. The principal sources of information regarding Cartier, are three *Relations*, of doubtful authorship, but possibly by Cartier himself. (*Cf.* Winsor, *Narr. and Crit. Hist. of America*, iv, 63-4.)

The many failures of the sixteenth century paved
the way for successes early in the seventeenth. It
was a troubled epoch in Europe. Germany was
on the verge of a long civil war. From England men,
impatient of an ecclesiastical system which they abhorred, went forth to found a New England in fervent
faith and intolerant zeal. Spain had absorbed Portugal, but was no longer an aggressive colonizing power,
when France, under her strong King, Henry IV., and
with acute religious strife for the time quieted, took
up colonial plans once more.

New France was founded several years before
New England, and its most conspicuous pioneer was
Samuel de Champlain. Above everything else Champlain was a traveller and an explorer. In early life he
had been in the West Indies, and in 1603 he
1603. made his first voyage to the St. Lawrence in
connection with the Canadian fur trade, which, even
then, had been for many years profitable, though
there were as yet no French settlements in Canada.
In the following year we find Champlain in what pro-
mised to be a more hospitable region. The
1604. present Province of Nova Scotia, called Acadia
by the French, was the scene of the first permanent
French settlement in North America. A French
nobleman, de Monts, secured leave to colonize
Acadia, together with a monopoly of the fur trade,
and in 1604 Champlain accompanied him to this
new field of labour. After some preliminary essays,
the banks of the fine inlet, now known as the
1605. Annapolis Basin, were selected for the first agricultural colony in North America. Noblemen,
workmen, missionaries, all found their way to Port
Royal, and it was here that the Jesuits made the first
of their aggressive efforts in North America, which
have so dramatic a history.

The prosperity of Port Royal was short lived. Far
down the Atlantic coast the English, too, had gained at

last a foothold on the American continent. The colony
1607. of Virginia, begun in 1607, developed rapidly,
and soon the English colonists learned of a
rival French settlement in the North. European
States were then as jealous of each other in America,
as they are now regarding Africa. In 1614, the
1614. Virginia colony sent an expedition northward to dislodge the French intruders, and
the leader Argall captured and burned Port Royal
and carried off all of the French upon whom he could
lay hands. A few found refuge in the woods, and ere
long they returned, sadly crippled, to take up once more
the work at Port Royal, and to become the nucleus of
the Acadian people, destined to have a tragic history in
1632. later times. Again, in 1628, the English seized
Acadia. The Scottish people were anxious to
begin colonizing work, and Sir William Alexander, afterwards Earl of Stirling, had already been given ill-defined
grants in those regions. To encourage settlement, an
order of Nova Scotia baronets had been established.
Alexander made some attempt at colonization, and the
1632. Acadian people of a later time contained a
certain Scottish intermixture, but in 1632,
Acadia was handed back to France. Colonization
1710. began anew, until in 1710, after a chequered
history, the British finally seized Port Royal,
and Nova Scotia passed permanently into their
possession.

Champlain, although interested in the preliminary
efforts, had not cast in his lot with Acadia. He turned
1608. again to the St. Lawrence, and in 1608, he
selected for his first settlement a spot destined
to be the most famous in the history of the North
American continent. On the strand of the River St.
Lawrence, with the high rock towering above him, he
established himself with a small company, and Quebec
soon became the centre of political and commercial
life in Canada. Champlain's heart was in exploration.

The mystery of the unknown interior haunted him.
With great difficulty he made his way by the Ottawa
River and Lake Nipissing to Lake Huron, one of the
vast bodies of fresh water in the interior. First of

1615.   Europeans he passed through the country near
the spot where Toronto now stands. He
reached Lake Ontario, crossed to its southern shore,
and with the Huron Indians, his allies, entered the
country of their enemy, the Indian confederacy
known as the Iroquois. Champlain took part in
this Indian warfare—an ominous beginning of a
struggle between the French and the Iroquois, which
was to last for a hundred years.

1628.   In 1628, England was at war with France.
A new trading company had just been organized by Richelieu himself—the Company of One
Hundred Associates—to unite all the fur trading mon-
opolists, and to govern Canada as the feudal vassal
of the King of France. The company came into
existence, however, at an inauspicious time. The
English under Sir David Kirk, appeared before Quebec.

1629.   Champlain was helpless and was carried off a
prisoner to England. For three years the
English flag floated over Quebec, and when the
country was restored to France in 1632, Champlain's
course was nearly run. He came back as Governor
under the new company in 1632, and died in 1635—a
truly great man who did his duty in a narrow and
difficult sphere of action. When he died, " New France "
meant a few settlers at Port Royal, two or three tiny
trading posts on the St. Lawrence, and less than two
hundred Frenchmen holding the rock at Quebec.*

With Champlain's death, the civil power ceases to

---

*Parkman in *Pioneers of France*, already cited, gives an adequate
account of Champlain, marred somewhat by laboured picturesqueness.
Champlain's narrative of his voyages, has been republished (6 Vols.
Quebec, 1870); an English translation has been published by the
Prince Society (Boston, 1878-82, 3 Vols.).

be the most important in Canada, and for thirty years
New France becomes a mission in which the dominant
interest is that of evangelizing the aborigines. Recollet
friars had been in the country with Champlain from
the beginning. In 1625, however, the Jesuits, disap-
pointed in their Acadian sphere of labour, turned to
Canada, and soon were sending out some of their most
heroic members to the perilous work among the Indians.
Their principal house was at Quebec, and it was to the
Huron mission that they devoted chief atten-
1625 to tion. From 1625 to 1649, the route by Lake
1649.
Ontario being closed by the hostile Iroquois,
the missionaries made the long and toilsome journey
by way of the River Ottawa to the Georgian Bay.
The Hurons, inhabiting the region north of Toronto
between Lake Simcoe and Lake Huron, treated them
with suspicion, and the Iroquois hated and pursued
both Hurons and French. It was soon apparent that
superior force was with the Iroquois. They closed in
on the Hurons and their Jesuit friends, and torture
and massacre were the fate of those who fell into the
hands of the savage conquerors. In 1649, the Huron
settlements and the French missions alike disappeared.
Among the Jesuit martyrs, Brébeuf and Lallemant, are
preëminent for the heroism with which they met their
awful fate. The French withdrew from the Huron
country, carrying with them the remnant of that
unfortunate people whose descendants may still be
seen near Quebec. The Jesuits soon began mission
work among the Iroquois themselves and there were
also other mission workers in the field. Montreal,
founded in 1640, became the headquarters of the
Sulpicians, the rivals of the Jesuits.*

After Richelieu's death in 1642, no leading statesman

---

* Parkman's *Jesuits in North America*, is a dramatic narrative of the
Huron mission. De Rochemonteix in *Les Jésuites et la Nouvelle
France* (Paris, 1895, 3 Vols.), gives a less fascinating account, from
the Jesuit standpoint.

in France had taken up a vigorous colonial policy Mazarin cared nothing for Canada. When, however, Colbert in 1661 succeeded Mazarin, colonial plans were once more actively pursued. The supremacy of the missionary idea really came to an end in 1663 when Canada was made a royal province, with Governor and Intendant on the model established by Richelieu for the provinces of France. Soon the Company of One Hundred Associates surrendered its powers and a new plan was formed for developing the French colonies. The *Compagnie des Indes Occidentales* was given trading privileges in both the West Indies and Canada and in 1665 the Marquis de Tracy, Lieutenant General for the King in all his American possessions, visited Canada to plan for better things than had yet been accomplished. At this time came one of the finest regiments of France —that of Carignan-Salières—now, after valiant service against the Turk, to face the hostile Iroquois. Above all Canada needed peace and peace could be secured only by convincing the Iroquois of the power of France. De Tracy led a military expedition into the Iroquois country and burned some of their villages. For the first time the frightened Indians saw France in her strength and hastened to make terms; during nearly twenty years New France had the blessings of peace.

There were troublesome domestic problems. Count Frontenac became Governor in 1672 and is undoubtedly the most striking figure among all the French governors of Canada. He was trained in the imperious school of Louis XIV. and came out determined that he, as representing the royal authority in Canada, would be as supreme as was the King in France. In Bishop Laval, and his successor, Saint Vallier, Frontenac met men with equally strong views of the Church's rights. The disputes between them were often unedifying and petty but the real issue was

1663.

1665.

1866.

1672.

whether New France should be a state or a mission.
Colbert had planned to occupy the interior. Mar-
quette, Joliet, La Salle and others, were pressing
in to find what was there, to establish French
settlements and to extend French trade. Laval and
the missionaries had no colonizing plans. They
dreaded the influence of the Europeans upon the
natives, for they saw that European trade was destroy-
ing the aborigines. The taste of these tribes for
strong drink is well-known and of all European
commodities brandy was most in demand by the
Indians. By 1670 it had actually become the medium
of exchange in the interior of North America, and
disease and crime followed its use by the natives.
Laval would have prohibited the trade entirely. To
Frontenac, on the other hand, control of the Indian
trade and of the Indians was a necessity for New
France. The Dutch and the English were rivalling
the French traders with cheaper and better European
wares. Brandy the Indians would have, and the pro-
tests of the Church against the trade were met by the
retort that if the natives were driven to trading with
the Protestant English and Dutch they would be destroyed not only by brandy but by heresy. The Court
of France was harassed by appeals from the contending parties. For a time the Church triumphed and
1682-89. Frontenac was recalled, but he soon returned
to save the colony which, without his strong
influence over the Indian character, had drifted again
into war with the Iroquois.

It was under Frontenac that the English and French
first engaged seriously in the struggle for supremacy
that was to end in the expulsion of France from the
North American continent. Under the two last
Stuarts England had been practically the vassal of
France but after the Revolution in 1688 there was no
longer any check upon the jealous hatred with which
the two nations viewed each other. Sir John Seeley

in his *Expansion of England* has shown that their long conflict which lasted from 1688 to 1815 finds its most serious meaning in commercial rivalry in remoter parts of the world and especially in India and America. Frontenac, in Canada, was a watchful foe of English expansion. He tried to arouse the Indians against them. He planned to occupy for France the points of vantage upon the St. Lawrence, the Great Lakes, the Ohio and the Mississippi. He attacked the English colonies and, had he been supported from France, might have taken Boston and New York. The English were driven from Hudson Bay. Louis XIV., however, was apathetic about distant Canada and the influence of

1696. the Church was against Frontenac. In 1696, he was ordered, in effect, to abandon the posts

1697. in the interior. In 1697 France made peace with England and restored what she had won in the

1698. colonies. Frontenac died in 1698 having apparently failed. His policy, however, soon

1701. triumphed. His successor made peace in 1701 with the Iroquois who cease henceforth to be formidable; the posts in the interior instead of being abandoned were strengthened; immigration began anew and New France passed out of the mission stage forever.*

1713. The year 1713 was a dark one for French colonial development. Louis XIV., exhausted by a renewed war, made peace at Utrecht on humiliating terms. France surrendered her claims to Hudson Bay, to Newfoundland, and to Nova Scotia, which then passed finally into the possession of Great Britain. France retained, however, the Island of Cape Breton and also what is now Prince Edward Island. Further struggle was inevitable and to be ready for it she built on Cape Breton at huge expense, Louisbourg, a fortress that

---

* Parkman's *Frontenac and New France Under Louis XIV.* is interesting. M. Lorin's *Comte de Frontenac* (Paris, 1895), enters more fully into Frontenac's colonizing plans.

1720 to 1745. soon became the strongest military post in North America. It commanded the St. Lawrence, menaced the English colonies, and served as a port of refuge for French commerce in the North Atlantic. At the same time France was grasping the whole interior of the continent. A French fort stood sentinel at the mouth of the Mississippi. By 1720.

1720 there were French traders as far west as the present Province of Manitoba and twenty years later the Canadian La Vérendrye had made 1743. his way across the continent to the Rocky Mountains. On the Ohio and Mississippi a line of French posts was planned to complete a continuous chain from Quebec to Louisiana.

Under Walpole's long ministry Great Britain's policy was peace. The year 1744, however, saw her 1744. once more in conflict with France. The first striking exploit of the renewed contest was the capture of France's great fortress of Louisbourg in 1745 1745. by New England militia forces aided, however, by a British fleet under Warren. England, by this time, had won command of the seas and France was unable to regain by force of arms what she ultimately secured by negotiation. A hollow peace, made in 1748, restored Cape Breton to France. Great Britain determined, however, to hold Nova Scotia, and she founded in 1749 the city of Halifax, which continues to be her strongest fortress in America. Great inducements were offered to colonists. The French, in alarm, disputed the English advance northward into what is now New Brunswick and stirred up the Acadians against their English conquerors. In 1750 there was fight-1750. ing on the Nova Scotia frontier and in other quarters France and England stood arrayed against each other in a time of nominal peace. The French were establishing military posts on the Ohio and the colonies of Virginia and Pennsylvania were profoundly interested in checking a French advance that should prevent English expansion westward. George Wash-

1754. ington, a young Virginia colonel of militia, sent out with a party of observation, came into conflict with a French force near the site of the present city of Pittsburgh. There was bloodshed, and Great Britain, still in a time of nominal peace, sent out to America a strong force of regular soldiers under Braddock to drive the French from the Ohio

1755. valley. Advancing through Virginia to attack the French fort, called DuQuesne after the Governor of Canada, Braddock fell into an ambush and was killed with many of his force.

The declaration of war, inevitable after such events, came in 1756. In Nova Scotia the English were alarmed lest the Acadians should join the side of France and the Governor, Lawrence, determined to send

1756. these unfortunate people out of the country. They were seized at various points and crowded into transports. Cargoes of the Acadians were landed at the seaports of the English colonies. Members of the same family were in some cases separated and unable to get any knowledge of each other's whereabouts, and the English colonies gave a cold and sometimes a cruel reception to their involuntary visitors. Ultimately, however, a large number of the deported Acadians found their way back to Nova Scotia or to Canada and their descendants in the Dominion now number nearly a quarter of a million.

In the war Great Britain at first made little pro-

1757. gress. In 1757, however, Pitt, though not Prime Minister, became the moving spirit of the administration, and infused new vigour into the

1758. American war. Amherst and Boscawen captured Louisbourg in 1758, and the French stronghold was destroyed. The English closed in on Canada. Wolfe in command of an army, and Saunders in command of a fleet, appeared before Quebec. In other quarters the British slowly advanced. The French

were forced to abandon the post on the Ohio that had
cost the English so much and Amherst was threaten-

1759.   ning Montreal from the south when, on a September day in 1759, Wolfe fought before Quebec
that brief battle in which both he and the French
leader Montcalm lost their lives. A few days later
the British flag waved over Quebec. Montreal soon

1763.   fell. The Treaty of Paris confirmed the British conquests and France lost her footing in
North America.*

In the minds of many, Great Britain won a doubtful
advantage in securing Canada, and it was soon to be
seen that the English colonies, never very pliant, would
be much less so when the menace of French attack was
finally removed. After the conclusion of the peace of
1763 British military officers, General Murray first, and
then General Sir Guy Carleton, ruled Canada. The re-
gret of the aborigines for France's loss was seen in the
cruel border warfare waged for a time against the English under the Indian Chief, Pontiac. English trading
adventurers came to seek their fortune in Canada and
it is not unnatural that the aristocratic governors
should have turned against these *bourgeois* of their
own race to the French *noblesse* still remaining
in the country. The conqueror's treatment of the
conquered was mild in the extreme and the prosperity
of Canada was soon greater than it had ever been
before. The people were free from military service,
and, for a time, from the Church's tithe and their
feudal obligations to the seigneurs. It was necessary to seek parliamentary sanction for the govern-

1774.   ment of Canada and in 1774 the British
Parliament passed its first enactment establishing a colony. It was obviously impossible to

*Parkman's *Half Century of Conflicts*, (Boston, 1892, 2 vols.), deals
with the fifty years prior to the outbreak of the Seven Years'
War. His *Montcalm and Wolfe*, (Boston, 1884, 2 vols.), is a brilliant
narrative of the British Conquest of Canada.

10

give control of the administration to the newly-conquered French. On the one hand, their feelings were respected by the re-establishment of the French civil law, although English criminal law was introduced, by restoring to the *noblesse* their feudal powers, and to the Roman Catholic Church the privileges she had formerly enjoyed. On the other hand, however, a despotic executive, entirely in the control of the Governor and his council, ruled the country. The Quebec Act excited dissatisfaction in three quarters. The British settlers in Canada, engaged chiefly in trade, were displeased at being placed on the same footing as the conquered French, with no voice in the government; the French *habitant*, to his disgust, saw himself once more subjected to the priest's tithe and to the feudal rights of the seigneurs; the neighbouring English colonies, now on the verge of revolt, saw Catholicism, which they abhorred, given peculiar privileges in Canada, a despotic government set up, and their own expansion westward checked by the inclusion of a vast western territory within the limits of the new " Province of Quebec."*

1775.   One of the first attempts of the revolted colonies, when the revolutionary war broke out, was to draw Canada into a revolt that should then include nearly the whole of North America. The troops of Congress invaded Canada and counted upon the French to join in throwing off the British yoke. The peasantry, with new aspirations for freedom and chafing under the revived feudalism imposed by the Quebec Act, showed a disposition to revolt. The *noblesse* and the Church appreciated the indulgent rule of the conqueror and stood loyal to Great Britain.

*Kingsford's *History of Canada*, vols. v-viii. (Toronto, 1892-95), covers the period from 1763 to 1815. Parkman's work does not extend beyond the British Conquest. Coffin, *The Province of Quebec and the American Revolution*, (Madison, Wis., 1896), discusses exhaustively the political effects of the Quebec Act.

The troops of Congress captured Montreal, and Generals Montgomery and Benedict Arnold laid seige to Quebec in the autumn of 1775. For a time it seemed as if Canada must fall. General Carleton, however, held out bravely against those whom he scorned as rebels and traitors. Montgomery was killed while leading a night attack, and ultimately Arnold raised the seige and the revolutionary troops retired from Canada. When, later, the French joined the colonies against Great Britain, and appealed to 1778. the national sentiment in Canada, Washington quietly discouraged a movement that might have resulted in giving France once more a footing in North America.

The successful revolt of the colonies profoundly affected the future of Canada. Many colonists had continued loyal to Great Britain. It is not to be wondered that, at a time when fierce passions were aroused, these loyalists should have been treated harshly. They were loaded with political disabilities and treated as social outcasts. Many suffered cruel personal indignities, banishment, and some even death. The property of loyalists was confiscated on an extensive scale, and all this was done without the judicial enquiry that should give the accused an opportunity of defence. When terms of peace were negotiated, Great Britain pleaded for mild treatment of those who had remained loyal to her. Upon this point, however, the representatives of the United States were inexorable. They declared that the federal government, which they represented, had no jurisdiction in the matter, and that each state had the right to determine its own policy in regard to the loyalists. It only remained for Great Britain to deal generously with those who had suffered in her behalf. Many were given Government posts or pensioned, and a sum of about $16,000,000 was distributed in compensation for losses. Thousands of the loyalists flocked

into Canada and were given free land.   Accurate
statistics are not available but the total influx did not
fall far short of one hundred thousand, and the present
population of Nova Scotia, of New Brunswick and of
Ontario is largely descended from them.   What is now
western Canada was then a part of the Province of
Quebec.   The newly arrived English settlers destroyed
its exclusively French character and they ere long
outnumbered the King's French subjects.*

Since the Conquest, government in Canada had been
necessarily despotic in type.   The loyalists, who now
came, had enjoyed, however, almost complete political
liberty in the English colonies and soon began to chafe
under absolutism.   Nova Scotia, the colony later known
as Prince Edward Island, New Brunswick, and the Province of Quebec, were all under separate Governments.
In the United States a new federal constitution was
being established and some discerning minds saw in 1789
that the British provinces should likewise be federated
in one state.   Not for eighty years, however, was this
to be brought about.   The movement, meanwhile, was
towards increased separation.   The Province of Quebec
was divided and in 1791 was replaced by
1791.   Lower Canada, of which the population was
chiefly French, and Upper Canada, peopled mainly by
Loyalists.   Each of the new provinces was given a
legislature which controlled taxation and legislation but,
like the Prussian Landtag and the German Reichstag
of the present day, had no control of the Executive
Government.   In the mother country Parliament had
already secured full control of the ministry, but
under the colonial theory of the time, the legislatures of
the colonies might not be entrusted with similar discretion in their own affairs.

---

*Cf. Sabine, *Loyalists of the American Revolution* (Boston, 1864, 2
vols.), and Ryerson, *The Loyalists of America and Their Times*,
Toronto, 1880, 2 vols.)

1792. The year in which this new scheme of Government was established in Canada saw Great Britain drawn into the long course of war brought on by the French Revolution. From 1792 to 1815, she was wholly absorbed in a momentous European conflict. When Napoleon rose to power he had plans for striking her in America, but his weakness on the sea made effective action in this direction impossible, though many of the French in Canada saw with sympathy the growth of a great French Empire. Canada meanwhile developed but slowly, for the European contest absorbed energies that might otherwise have led to emigration. There was, however, a considerable immigration from the United States to supplement the earlier Loyalist movement. These new settlers came to escape taxation, and to secure new and rich lands then easily obtained, especially in Upper Canada, and their republican convictions helped to cause the belief in the United States that Canada was ready to cast off allegiance to Great Britain. The two nations again drifted into war. The attempts of Napoleon to destroy Great Britain's commerce with continental Europe and the British retaliatory blockade of the ports of France and her allies even against neutrals, caused great irritation in the United States. This feeling was concentrated against Great Britain by her claimed right to stop and search American ships for British deserters, who were then very numerous in the naval service of the United States. The New England States wished to avoid a conflict, but elsewhere the war cry was potent and in 1812 the struggle began,

1812. although Great Britain had already removed the restrictions on trade that had caused the chief annoyance.

The war of 1812-14 is the only great contest in which Canada has been involved since there was any considerable English settlement west of Quebec, and Canadian patriotism still dwells with pride upon a

drama in which it played a creditable part. From
Lake Superior to Nova Scotia there was an intermit-
tent struggle but the chief centre of the war was on
the frontier formed by the River Niagara.

1812. In 1812, General Brock was in command of
the forces in Upper Canada. He won a signal success
by the invasion of American territory and the

August, capture of Detroit. including the American
1812. general, Hull, and his force. A few months
later Brock fell, repelling an invasion of Canada

Oct. 13th, near Niagara Falls, and a noble monument on
1812. Queenston Heights now marks the scene of his
gallant death. The invading force was driven back
for the time, but in the second year of the

1813. war the Americans secured naval control of
Lakes Ontario and Erie. Toronto, then a tiny capital,
was taken and its public buildings burned, an event
grimly revenged in the following year, when the
British burned the public buildings at Wash-

1814. ington. Niagara, the former capital of Upper
Canada, was also burned by the Americans. The
British, under General Procter, were driven from
Detroit back into Upper Canada, and suffered a dis-
astrous defeat at Moraviantown by the American
general, Harrison, subsequently President, and

October, the grandfather of a President, of the United
1813. States. Yet, notwithstanding these reverses,
the British and Canadian forces under General Drum-
mond, small though it was, drove back the invaders
before the end of the season, and occupied for a time
and ravaged the American frontier along the whole
length of the Niagara River. In 1814, the struggle

1814. increased in vehemence. At Lundy's Lane,
near Niagara Falls, a fierce struggle lasted

July 25th, far into the night, and at its close the British
1814. remained in possession of the field. A British
squadron had meanwhile secured control of
Lake Ontario and finally, in November, the Americans withdrew from Upper Canada.

In Lower Canada, the commander of the British forces was Sir George Prevost, the Governor-General, a singularly unfortunate, if not an entirely incompetent, leader. The French Canadians fought well against the invaders. A junction of two American forces before Montreal was prevented by victories at Châteaugay, near Montreal, and at Chrysler's Farm, on the St. Lawrence. To Lower Canada in 1814, Great Britain sent out powerful reinforcements, among them many veterans who had fought with Wellington in the Peninsula. A combined naval and military attack on Plattsburg, an American post on Lake Champlain, failed utterly, and ten thousand troops, composed of the flower of the British army, retreated before an inferior force.

Oct. and Nov., 1813.

September, 1814.

Events, meanwhile, strengthened the hands of Great Britain, for Bonaparte had fallen. The chief causes of the irritation that had led to the war with the United States had already been removed, and in New England bitter opposition to the continuance of the struggle was growing. Finally, in December, 1814, a fratricidal war, purposeless in origin and almost fruitless in result, came to an end.

December, 1814.

The close of the great European war brought about new economic conditions in the British Isles and between 1820 and 1837 there was an extensive emigration to Canada. It was now that the system of government established in 1792 was first seriously tested and it was found to be deplorably wanting. Lord Durham, when urging in 1838 that Great Britain should change her policy in regard to Canada, added "The experiment of keeping colonies and governing them well ought at least to have a trial"—a biting comment upon two hundred years of colonial activity. The intentions of the Home Government were admirable. The fault was in the system and one of its great defects was the want of

1820 to 1837.

continuity in the policy of the colonial office. The
ministers were changing ceaselessly. In the ten years
before the outbreak of rebellion in 1837, there were no
less than eight colonial ministers, each with a policy of
his own. Canadian affairs were dependent upon party
exigencies in London, and the needs of Canada were
not understood where its affairs were controlled.

The problems to be faced in the two provinces were
essentially different. In Upper Canada public affairs
were controlled by an official and professional class
who stood together, surrounded the Governor, and
ruled often in defiance of the wishes of the people's
representatives in the legislature. This class had the
faults that irresponsible power begets. The central
bureaucracy controlled the affairs of all sections of the
country. Notwithstanding the high character of some
of its members it acted selfishly and even corruptly.
It monopolized the lucrative posts in the public service
and in the Anglican church. Huge grants of land
were made to influential persons and legitimate settlement was checked by land monopoly. Great public
works were undertaken to connect Upper Canada with
the seaboard and were corruptly administered. Education was neglected. Municipal liberties and healthy
local initiative were checked. An Act of the Imperial
Parliament, providing that one-seventh of the public
lands should be set aside for the use of a "Protestant"
clergy, was interpreted as establishing the exclusive
claims of the Anglican church, though other Protestant
bodies represented a majority of the population.
Roman Catholics were excluded from all share in the
work of government. Ardently British in tone, the
members of the ruling class yet looked with jealousy
upon immigrants from the mother-country and the
franchise was given to them on illiberal terms. Professional men coming from Great Britain were required
to serve a long apprenticeship before they were allowed
to practise in the country. Immigration from the

United States, which had been extensive at an earlier period, was also discouraged by the refusal to allow American citizens to hold land, and Lord Durham complained in 1838 that, instead of attracting settlers, Upper Canada was being depopulated by emigration to the United States. The reforming party in the Legislative Assembly agitated and protested against the policy of the Government but even when it controlled the Chamber the executive could defy them, and at last in 1837 some of the more reckless of the Reformers, headed by William Lyon Mackenzie, appealed to arms.

In Lower Canada there was a similar conflict between the elected assembly and the Government but otherwise the problems were quite different. A war of races was going on in Lower Canada. During the earlier years of British supremacy there had been intercourse and intermarriage between the English and the French. As the English increased in number jealousy between the two races also developed and a bitter political warfare began. The masses of the people were French and for them there was the double grievance of subjection to an alien race and to the oppressive feudalism that the British had re-established. Many of the seigneuries had been purchased by Englishmen who vigorously enforced their feudal rights. Commerce was also chiefly in the hands of the English. The English minority, democratic and progressive in principle, yet leagued itself with the despotic executive. The Catholic church and the French *noblesse*, secure in their privileges under the existing system, were opposed to any violent change. The social separation of the two races was complete. Religious dissension were not acute, largely because of the complete isolation of the one side from the other. The only protest which the people could make against the actions of the Government was through the elected legislature and it was powerless to effect a change. It often embar-

rassed the Government by refusing consent to the
most necessary enactments and by denying supplies.
Juries refused to convict obviously guilty persons if
the race question was involved.   The breach steadily
widened and finally, in 1837, in Lower as in Upper
Canada, the discontented made a rash appeal to arms.
French Canadian patriots even dreamed of defying the
power of Great Britain and of founding a French
Republic on the banks of the St. Lawrence.

In Upper Canada the struggle was almost bloodless.
An unthinking intrusion by the loyal party upon the
territory of the United States nearly provoked a
wider conflict. The steamer *Caroline*, engaged in
the service of those in rebellion, was moored on the
American side of the River Niagara and a Canadian
force cut her out, set her on fire and sent her over the
torrent. The event caused a fierce agitation in the
United States against Great Britain and thus became
the most conspicuous incident of the rebellion in Upper
Canada. In Lower Canada there were some bloody skir-
mishes at St. Denis, St. Charles and St. Eustache, but
only a few thousand badly armed peasantry joined the
revolt. The political leader Papineau fled to the United
States and his colleague Nelson was soon captured.
The struggle was not fruitless, however, for it attracted
the attention of the mother-country to existing evils
in Canada. The other British provinces in North
America had their own grievances.   In Prince Edward
Island unwise distribution had left the land largely in
the possession of absentee proprietors ; in Nova Scotia
and New Brunswick there was the usual struggle between the Government and the representatives of the
people. The Home Government took decisive action
in 1838 when Lord Durham, one of the leading
liberal statesmen of the day, was sent out as
High Commissioner and Governor-General of all the
British Provinces. His conclusions are embodied in a
masterly report in which he declared that full respon-

1838.

sible government must be given to all the North American colonies and that, to insure English supremacy, the French Province of Lower Canada must be united with the English Province of Upper Canada. He expressed a hope for the ultimate union of all the colonies and declared that only when they thus became a great state would the Home Government cease to intervene in their affairs.*

A new Constitution for Canada resulted from the events of 1837-38. In 1841, by an Imperial Act of Parliament, Upper and Lower Canada were united into one Province with one Legislature. The union of two peoples in one Legislature, Lord Durham had hoped, would give permanent supremacy to the English. In fact, it did not do so. French Canada, known usually after 1841 as Canada East, was given equal representation with Canada West.

1847. Not until 1847 was responsible Government secured, and with responsible Government came a new difficulty. The Government of the day must command the support of both the French and the English elements and in practice it was found necessary to have leaders for each section. So complete was still the political separation of French and English Canada that each leader was required to have a majority in his own province. The difficulty of carrying on a Government which was obliged to have a French and Catholic majority in Canada East, and an English and Protestant majority in Canada West, is obvious. It sometimes happened that a Government with an absolute majority in the House was compelled to resign because its supporters from one of the provinces were in a minority. For more than twenty years Canada

*Lord Durham's *Report on the Affairs of British North America* (London, 1839), contains an admirable account of the causes of the disputes in the British Provinces. The Report is understood to have been written by Charles Buller.

staggered under this impossible system, and then the deadlock of Government made all parties see the necessity of a radical change.

During this period, however, some of the troublesome political questions that vexed the country had been settled. The Government obeyed the wishes of a majority of the people. With the consent of the Imperial Parliament, the supremacy of the Anglican church was brought to an end by secularizing the state university in Canada West, and by devoting the greater part of the disputed "Clergy Reserves" land to general educational purposes. At the same time feudalism was abolished in Canada East, and the *habitant* was freed at last from the grievance of generations. There was a renewed flow of immigration, and an advantageous reciprocity treaty with the United States, made through the energy and ability of Lord Elgin, Governor of Canada, helped to promote rapid commercial expansion.

1854.

1854.

In 1844, after the dissolution of the first Canadian Parliament since the Union, Mr. John A. Macdonald, a young barrister, had been elected member for Kingston. In 1847, he took office in the Government of the day, and almost continuously from that time until his death in 1891, he was the most conspicuous figure in the political life of Canada. Brilliant, adroit, fond of power, careless of the means or agents he used to effect his ends, passionately devoted to what he conceived to be the welfare of Canada, and to British connection, Macdonald combined in an extraordinary manner the qualities both of the statesman and of the politician. The most prominent of his opponents was Mr. George Brown, a Scot of strong religious convictions and determined will and fiercely opposed to religious and class privileges. Brown, who hated Roman Catholicism, wished to secure English supremacy by giving Canada West a larger representation in the legislature on account of its increased population. The necessities of the political situation at last brought Macdonald and

1864. Brown into alliance. When they united in 1864, four ministries had been defeated within three years, and cabinets were rapidly becoming as unstable in Canada as they have been in France under the Third Republic. It happened that in 1864, a movement was on foot in Nova Scotia, New Brunswick and Prince Edward Island, to promote economy by uniting the Governments. Delegates from Canada were present at a conference of representatives of the three provinces, held at Charlottetown, P. E. I., in September, 1864. A year later this meeting had ripened

1865. into a council of delegates from the British North American provinces, held at Quebec, to discuss the question of confederation.

When the Canadian Confederation was outlined the United States had just ended a great civil war. Its main issue related to the claim by individual States to be independent commonwealths and to have the right to withdraw at discretion from a Confederation which they had voluntarily entered. Macdonald sought to avoid in Canada the weaknesses that time had revealed in the American system. His ideal was not a federation, but a legislative union similar to that of the United Kingdom, and he was firm in insisting that individual provinces should have strictly limited powers. He was the guiding spirit of the Quebec Conference, and it was under his influence that it adopted proposals to establish a " Kingdom of Canada," which should be an auxiliary kingdom under British sovereignty. The local legislatures in all the provinces were to be retained, with powers greatly curtailed. It was necessary to go to London for the necessary legislation by the Imperial Parliament and in London the Canadian plan was modified in some respects. At the time there was danger of collision between Great Britain and the United States, owing to causes springing from the civil war. Anxious to avoid any cause of offence to a

republican people, Lord Derby insisted that Canada
should be not a " Kingdom " but a ' Dominion," and
finally the Imperial Parliament passed the British
North America Act under which the Canadian Confederation was established.   Entrance to the
Confederation was to be voluntary on the part
of the provinces.   Within a few years all had
entered except Newfoundland.   The territory of Canada
extended in 1871 from the Atlantic to the Pacific, for
the remote Province of British Columbia then entered
the Confederation on the condition that a railway
should be built across the continent within Canadian
territory.   Delay in fulfilling this condition caused
some irritation in the Pacific province, speedily removed, however, by the completion of the Canadian
Pacific Railway.*

1867
to
1873.

Only with the establishment of the Dominion of
Canada did the British provinces in North America
find a workable political system.   The experience of
thirty years has vindicated the wisdom with which
the Constitution was planned.   The principal political
issues since Confederation have related to internal
development and to external trade.   Macdonald's active mind conceived large enterprises which he pressed
on to rapid completion.   His opponents in the Liberal
party would have proceeded less rapidly in carrying
out great public works.   In commercial matters Macdonald advocated a high protective tariff—a " National
Policy " as he termed it—while the Liberals favoured a
tariff for revenue mainly.   From 1878 to 1896 the party
of Protection triumphed.   It remains to be seen in what
degree the Liberal party, now in power, will move
towards Free Trade.   Already they have discriminated
in favour of the mother country.

---

*Mr. J. C. Dent's *The Last Forty Years* (Toronto, 1881, 2 Vols.), is
a popular history of recent events.   Pope's *Life of Sir John Macdonald* (London, 1894, 2 Vols.), is really a history of the times.   *Cf.* also
Mackenzie's *Life of George Brown* (Toronto, 1882).

No serious conflict has disturbed the peace of Canada since 1838. In 1866 when the United States were greatly irritated against Great Britain for causes arising out of the Civil War, their territory was made the basis of an insignificant Fenian attack upon Canada. In 1871, Canada assumed authority over the territory of the Hudson Bay Company, which included the present Province of Manitoba, and there was a slight revolt on the part of half-breed settlers to whom the Dominion seemed an alien State. A second rising of half-breeds followed in 1885, under the same leader, Louis Riel, but it was promptly suppressed by the Canadian militia.

Relatively to the United States, the growth of population in Canada has been slow. When we remember, however, that at the time of the British Conquest in 1763, the population of European descent in what is now Canada was scarcely 100,000, and that at present it is more than 5,000,000, the record is sufficiently striking. Canada has nearly twice as many inhabitants as had the thirteen colonies when they revolted against the mother-country. The Canadian people have developed a political system that provides in a remarkable degree for full personal, municipal, and national liberties; they have no serious grievances that cannot be remedied by themselves; they are contented and fairly prosperous. As DeTocqueville foretold, the equality of conditions in the new world has irresistibly led to democratic Government. His other prophecy that the higher types of European culture would develop only slowly if at all, amid this equality of conditions, has proved equally true. There has been, hitherto, little leisure or field in Canada for the extensive cultivation of literature and art. Yet in nearly every town and village of the English provinces, there is a public library; in intelligence and education the tillers of the soil in these provinces, will compare favourably with the peasantry

of any other country, while in the cities there are
almost no slums, and the poorer population is surprisingly well housed.

The political connection of Canada with Great Britain
has had one striking result.   People from other European countries, have preferred to emigrate to the United
States rather than to become the subjects of a rival
European State, by settling in Canada.   Immigration
has been in consequence almost entirely from the United
Kingdom, and the population is thus, in origin, vastly
different from the heterogeneous masses in the United
States.   Immigrants have for the most part already
been trained in the British school of political action,
and have learned self-control and moderation.   On the
other hand, Canada, like the United States, has a
serious race problem.   Side by side with 3,000,000
people of British origin, are nearly 2,000,000, French
in origin and in mental characteristics.   The federal
system has minimized the friction between races, and,
while all difficulties are not removed, Lord Durham's
statement that in Canada there were " two peoples
warring in the bosom of a single state," is no longer
true.

CHAPTER III.

An Outline of the Constitutional History and System of Government (1608-1897).

By J. G. Bourinot, C. M. G., F. R. S. C., Author of "How Canada is Governed," "The Story of Canada," and Other Works on the History and Government of the Dominion.

THE Dominion of Canada comprises a number of provincial divisions, whose political and constitutional histories had no direct connection until 1867, except so far as they were all subject to the supreme authority of England from the time when France gave up her project of establishing an empire in America. First of all, we have that great country in the valley of the St. Lawrence, which was originally called Canada under the French regime, and has now given its name to a confederation stretching from ocean to ocean. Then we have the Maritime Provinces which were formerly known as the ill-defined territory of Acadie and a section of which, the present Province of Nova Scotia, was organized as a British possession long before the cession of Old Canada. In addition to these old Acadian and Canadian divisions there is the great region of prairies and mountains, stretching from Lake Superior to the Pacific Ocean, for centuries the hunting ground of fur-traders, and only brought into the pale of civilization and organized government within the latter half of the present century.

As the most convenient method of dealing with my subject, I shall confine myself for the present to the political history of the large country generally known as Canada until 1867, and now divided into the Pro-

11

vinces of Ontario and Quebec. This history may be
properly divided into several periods, from the time
Champlain laid the foundation of the French colony on
the banks of the St. Lawrence, down to the establishment of the system of federation.

First of all, we have the period when France claimed
dominion over the extensive ill-defined territories
watered by the St. Lawrence and the Great Lakes, and
including the valleys of the Ohio and the Mississippi
Rivers. During this period, which lasted from 1608 to
1759-60, Canada was under the control for a number
of years of proprietary governments chartered by the
King to carry on trade in the country. By 1663,
however, Louis XIV. decided, under the advice of the
eminent statesman Colbert, to take the government of
Canada into his own hands. The plans of 1663 were
fully carried out in 1674. The governor of Canada
and the intendant were to all intents and purposes, in
point of authority, the same officials who presided
over the affairs of a province of France. In Canada,
as in France, governors-general had only such powers
as were expressly given them by the King, who, jealous
of all authority in others, kept them rigidly in check.

The governor had command of the militia and
troops, and was nominally superior in authority to the
intendant, but in the course of time the latter became
virtually the most influential officer in the colony, and
even presided at the council board. He had the right
to report directly to the King on colonial affairs, had
large civil, commercial and maritime jurisdiction, and
could issue legal ordinances on his own responsibility.
Associated with the governor and intendant, was a
council, who exercised legislative and judicial powers,
and was a Court of Appeal from the judicial function-
aries at Quebec, Montreal and Three Rivers, the
principal towns of the three districts into which the
country was divided for the administration of justice
in accordance with the *Coutume de Paris*. The bishop

was a member of the council. The Roman Catholic church, from the very first settlement of Canada, was fostered by express provisions in the charters of the incorporated commercial companies. When the King assumed the government, the bishop and his clergy continued to increase their power and wealth, and by the time of the conquest the largest landed proprietors, and in many respects the wealthiest, were the church and its communities. The seigniory soon gave way to the parish of the church, as a district for local as well as for ecclesiastical purposes. Tithes were imposed and regulated by the government, and as the country became more populous the church grew in strength and riches.

The King and the council of state in France kept a strict supervision over the government of the colony. We look in vain for evidence of popular freedom or material prosperity during these times. The government was autocratic and illiberal, and practically for many years in the hands of the intendant. Public meetings were steadily repressed and no system of municipal government was ever established. It is not strange then, that the *habitants* of the seigniories, as well as the residents in the towns, lived for the most part a sluggish existence without any knowledge of, or interest in the affairs of the colony, which were managed for them without their consent or control, even in cases of the most insignificant matters.

We must now come to the Second Period in our political history, which dates from the capitulation of Quebec and of Montreal in 1759-1760. It may be considered, for the purposes of historical convenience, to be the transition stage from the conquest until the granting of representative institutions in 1791. During this transition period it is interesting to notice the signs that the French Canadian leaders gave from time to time of their comprehension of the principles of self-government. Several political facts require brief

mention in this connection. From 1760 to 1763 when
Canada was finally ceded to Great Britain by the
Treaty of Paris, there was a military government as a
necessary consequence of the unsettled condition of
things. Then King George III. issued his famous
proclamation of 1763, and by virtue of the royal prerogative, established a system of government for
Canada. The people were to have the right to elect
representatives to an Assembly, but the time was not
yet ripe for so large a measure of political liberty, if
indeed it had been possible for them to do so under the
instructions to the governor-general which required all
persons holding office or elected to an assembly to take
oaths against transubstantiation and the supremacy of
the Pope. This proclamation created a great deal of
dissatisfaction, not only for the reason just given, but
on account of its loose reference to the system of laws
that should prevail in the conquered country. In
1774, the Parliament of Great Britain intervened and
passed the Act giving the first constitution to Canada,
generally known as the Quebec Act. The Act established a legislative council nominated by the Crown,
and the project of an assembly was indefinitely post-
poned. The French Canadians were quite content, for
the time being, with a system which brought some of
their leading men into the new legislative body. The
Act removed the disabilities under which the French
Canadians, as Roman Catholics, were heretofore placed,
and guaranteed them full freedom of worship. The
old French law was restored in all matters of controversy relating to property and civil rights. The
criminal law of England, however, was to prevail
throughout the country.

While this Act continued in force various causes
were at work in the direction of the extension of popular government. The most important historical fact
of the period was the coming into British North
America of some fifty thousand persons, known as

United Empire Loyalists, who decided not to remain in the old Thirteen Colonies when these forswore their allegiance to the King of England. This body of sturdy, resolute and intelligent men, united by high principles and the most unselfish motives, laid the foundations of the Provinces now known as New Brunswick and Ontario, and settled a considerable portion of Nova Scotia.

In view of the rapidly increasing English population of Canada and of the difficulties that were constantly arising between the two races,—difficulties increased by the fact that the two systems of law were constantly clashing and the whole system of justice was consequently very unsatisfactorily administered, —the British Government considered it the wisest policy to interfere again and form two separate provinces, in which the two races could work out their own future, as far as practicable, apart from each other. The Constitutional Act of 1791 was the beginning of the Third Period in the political history of Canada, which lasted for half a century. This Act extended the political liberties of the people in the two Provinces of Upper and Lower Canada—now Ontario and Quebec—organized under the Act, since it gave them a complete legislature, composed of a governor, a legislative council nominated by the Crown, and an assembly elected by the people on a limited franchise. The Constitution of 1791, though giving many concessions and privileges to the two provinces, had an inherent weakness, since it professed to be an imitation of the British system, but failed in that very essential principle which the experience of England has proved is absolutely necessary to harmonize the several branches of government ; that is, the responsibility of the executive to Parliament, or more strictly speaking to the assembly elected by the people. The English representatives in the Province of Upper Canada soon recognized the value of this all important principle of parliamentary government according as they had experience of the practical operation of the system actually in vogue; but it is an admitted fact that the French Canadian leaders in the assembly never appreciated the constitutional system of England in its full significance. Their grievances, as fully enumerated in the famous resolutions of 1834, were numerous, but their principal remedy was always an elective legislative council. The conflict that existed during the last thirty years of this period was really a conflict between the two races in Lower Canada, where the French and elective element predominated in the assembly, and the English and official or ruling element in the legislative council. The executive government and legislative council, both nominated by the Crown, were virtually the same body in those days. The ruling spirits in the one were the ruling spirits in the other. The English speaking people were those rulers, who obstinately contested all the questions raised from time to time by the popular or French party in the assembly. In this contest of race, religion and politics, the passions of men became bitterly inflamed, and an impartial historian must deprecate the mistakes and faults that were committed on both sides.

In Upper Canada the political difficulties never assumed so formidable an aspect as in the French Canadian section. No difference of race could arise in the western provinces, and the question of the control of supplies and expenditures—the chief objects of contention in the lower province—gradually arranged itself more satisfactorily than in Lower Canada, but in the course of time there arose a contest between officialism and liberalism. An official class held within its control practically the government of Upper Canada. This class became known in the parlance of those days as the " family compact," not quite an accurate designation, since the ruling class had hardly any family connection, but there was just enough ground for the term to tickle the taste of the people for an epigrammatic phrase. The "clergy reserves" question grew out of the grant to the Protestant Church in Canada of large tracts of land by the Constitutional Act, and was long a dominant question in the contest of parties. The reformers, as the popular party called themselves, found in this question abundant material for exciting the jealousies of all the Protestant sects who wished to see the Church of England and the Church of Scotland deprived of the advantages which they alone derived from this valuable source of revenue.

The history of this period was history repeating itself: the contest of a popular assembly against prerogative, represented in this case by the governor and executive which owed no responsibility to the people's house. All the causes of difficulty in the two provinces were intensified by the demagoguism that is sure to prevail more or less in time of popular agitation, but the great peril all the while in Lower Canada arose from the hostility of the two races in the political arena as well as in all their social and public relations. The British government laboured to meet the wishes of the discontented people in a conciliatory spirit but they were too often ill-advised or in a quandary from the conflict of opinion. No doubt the governors on whom they naturally depended for advice were at times too much influenced by their advisers, who were always fighting with the people's representatives, and at last in the very nature of things made advocates of the popular party. Too generally they were military men, choleric, impatient of control, and better acquainted with the rules of the camp than the rules of constitutional government and sadly wanting in the tact and wisdom that should guide a ruler of a colony. The political discontent was at last fanned into an ill-advised rebellion in the two provinces, a rebellion which was promptly repressed by the prompt measures immediately taken by the authorities.

In Lower Canada the constitution was suspended and the government of the country from 1838-1841 was administered by the governor and a special council. The most important fact of this time was the mission of Lord Durham, a distinguished statesman, to inquire into the state of the country as governor-general and high-commissioner. His report was a remarkably fair summary of the causes of discontent and suggested remedies which recommend themselves to us in these days as replete with political wisdom. The final issue was the intervention of Parliament once more in the affairs of Canada and the passage of another Act providing for a very important constitutional change.

The proclamation of the Act of 1841 was the inauguration of the Fourth Period of political development which lasted until 1867. The discontent that existed in Canada for so many years had the effect, not of diminishing but of enlarging the political privileges of the Canadian people. The Imperial Government proved by this measure that they were desirous of meeting the wishes of the people for a larger grant of self-government.

So far from the Act of 1841, which united the Canadas, acting unfavourably to the French Canadian people—as their leaders anticipated—it gave them eventually a predominance in the councils of the country and prepared the way for the larger constitution of 1867 which has handed over to them the direct control of their own province, and afforded additional guarantees for the preservation of their language and institutions. French soon became again the official language by an amendment of the Union Act, and the clause providing for equality of representation proved a security when the upper province increased more largely in population than the French Canadian section. The Act was framed on the principle of giving full expansion to the capacity of the Canadians for local government, and was accompanied by instructions to

the Governor-General, Mr. Poulett Thomson, afterwards
Lord Sydenham, which laid the foundation of responsible government. It took several years to give full
effect to this leading principle of parliamentary government, chiefly on account of the obstinacy of Lord
Metcalfe during his term of office; but the legislature
and the executive asserted themselves determinedly,
and not long after the arrival in 1847 of Lord Elgin,
one of the ablest governors-general Canada has ever
had, the people enjoyed in its completeness that system
of the responsibility of the cabinet to Parliament without which our constitution would be unworkable.
More than that, the clergy reserves and other difficult
questions were settled and all the privileges for which
the people had been contending during a quarter of a
century and more, were conceded in accordance with
the liberal policy now laid down in England for the
administration of colonial affairs.

Municipal institutions of a liberal nature, especially
in the Province of Ontario, were established, and the
people of the provinces enabled to have that control of
their local affairs in the counties, townships, cities and
parishes, which is necessary to carry out public works
indispensable to the comfort, health and convenience
of the community, and to supplement the efforts made
by the legislature, from time to time, to provide for
the general education of the country: efforts especially
successful in the Province of Upper Canada, where
the universities, colleges and public schools are so
many admirable illustrations of energy and public
spirit. The civil service, which necessarily plays so
important a part in the administration of government,
was placed on a permanent basis, and has ever since
afforded a creditable contrast with the loose system
so long prevalent in the United States.

The Union Act of 1841 did its work, and the political
conditions of Canada again demanded another radical
change commensurate with the material and political

development of the country, and capable of removing
the difficulties that had arisen in the operation of the
Act of 1841. The claims of Upper Canada to larger
representation, equal to its increased population
since 1840, owing to the great immigration which
naturally sought a rich and fertile province, were
steadily resisted by the French Canadians as an un-
warrantable interference with the security guaranteed
them under the Act. This resistance gave rise to great
irritation in Upper Canada, where a powerful party
made representation by population their platform, and
government at last became practically impossible on
account of the close political divisions for years in the
assembly. The time had come for the accomplishment
of a great change, foreshadowed by many public men
in Canada: the union of the provinces of British North
America. The leaders of the different governments in
Canada and in the Maritime Provinces of Nova Scotia,
New Brunswick and Prince Edward Island, combined
with the leaders of the Opposition with the object of
carrying out this great measure. These provinces had
been in the enjoyment of representative institutions
since the latter half of the eighteenth century. Nova
Scotia was given an Assembly as early as 1758; New
Brunswick was made a province, with representative
government, in 1754 ; and Prince Edward Island, from
1769-1773. Responsible government, as in the case of
Canada, had been fully granted by 1850 to all these
provinces, whose public men were now fully conscious
of the necessity of a union and a larger sphere of political action. A convention of thirty-three representative
men from all the provinces was held in the autumn of
1864, in the historic city of Quebec, and after a deliberation of several weeks the result was the unanimous
adoption of a set of seventy-two resolutions embodying
the terms and conditions on which the provinces,
through their delegates, agreed to a federal union.
These resolutions had to be laid before the various

legislatures, and adopted in the shape of addresses to
the Queen, whose sanction was necessary to embody
the wishes of the provinces in an Imperial statute. In
the early part of 1867 the Imperial Parliament, without
a division, passed the statute known as the " British
North America Act, 1867," which united in the first
instance the Province of Canada, now divided into
Ontario and Quebec, with Nova Scotia and New
Brunswick, and made provisions for the coming in of
the other Provinces of Prince Edward Island, Newfoundland, British Columbia, and the admission of
Rupert's Land and the great North-West.

Between 1867 and 1873 the provinces just named,
with the exception of Newfoundland, which has per-
sistently remained out of the federation, became parts
of the Dominion, and the North-West Territory
was at last acquired on terms eminently satisfactory
to Canada and a new province of great promise, Manitoba, formed out of that immense region, with a complete system of parliamentary government.

In accordance with this constitution, Canada has
now control of the government of the vast territory
stretching from the Atlantic to the Pacific to the north
of the United States, and is subject only to the sovereignty of the Queen and the Parliament of Great
Britain in such matters as naturally fall under the jurisdiction of the supreme and absolute authority of the
sovereign state.

If we come to recapitulate the various constitutional
authorities which now govern the Dominion in its external and internal relations as a dependency of the
Crown, we find that they may be divided for general
purposes as follows :—

The Queen.

The Parliament of Great Britain.

The Judicial Committee of the Privy Council.

The Government of the Dominion.

The Government of the Provinces.

The Courts of Canada.

While Canada can legislate practically without limitation in all those matters which do not affect Imperial interests, yet sovereign power, in the legal sense of the phrase, rests with the government of Great Britain. Canada cannot of her own motion negotiate treaties with a foreign state, as that is a power only to be exercised by the sovereign authority of the empire. In accordance, however, with the policy pursued for many years towards self-governing dependencies—a policy now practically among the "conventions" of the constitution—it is usual for the Imperial Government to give all the necessary authority to distinguished Canadian statesmen to represent the Dominion interests in any conference or negotiations—the Fishery and Behring Sea questions, for example—affecting its commercial or territorial interests. The control over peace and war still necessarily remains under the direct and absolute direction of the Queen and her great council. The appointment of the Governor-General, without any interference on the part of the Canadian Government, rests absolutely with the Queen's Government. The same sovereign authority may "disallow" an Act passed by the Parliament of Canada which may be repugnant to any Imperial legislation on the same subject applying directly to the Dominion, or which may touch the relations of Great Britain with foreign powers, or otherwise seriously affect the interests of the Imperial State. The Judicial Committee of the Queen's Privy Council is the court of last resort for Canada as for all other parts of the British Empire, although that jurisdiction is only exercised within certain limitations consistent with the large measure of legal independence granted to the Dominion. As it is from the Parliament of Great Britain that Canada has derived her constitution, so it is only through the agency of the same sovereign authority that any amendment can be made to that instrument.

The preamble of the British North America Act, 1867, sets forth that the provinces are "federally united," with a constitution "similar in principle to that of the United Kingdom." The model taken by Canadian statesmen was almost necessarily that of the United States, the most perfect example of federation that the world has yet seen, though they endeavoured to avoid its weaknesses in certain essential respects. At the same time, in addition to the general character of the provincial organizations and distribution of powers, and other important features of a federal system, there are the methods of government, which are copies, exact copies in some respects, of the Parliamentary Government of England.

The various authorities under which the government of the Dominion is carried on may be defined as follows :—

1. The Queen, in whom is legally invested the executive authority ; in whose name all commissions to office run ; by whose authority Parliament is called together and dissolved ; and in whose name bills are assented to and reserved. She is represented for all purposes of government by a Governor-General, appointed by Her Majesty in Council and holding office during pleasure ; responsible to the Imperial Government as an Imperial Officer : having the right of pardon for all offences, but exercising this and all executive powers under the advice and consent of a responsible ministry.

2. A Ministry composed of thirteen or more members of a Privy Council ; having seats in the two Houses of Parliament ; holding office only whilst in a majority in the popular branch ; acting as a council of advice to the Governor-General ; responsible to Parliament for all legislation and administration.*

---

*The Ministers of the Cabinet are : President of Council, who presides over the meetings of the Cabinet, and is at present the Prime Minister ; Minister of Justice and Attorney-General of Canada,

3. A Senate composed of eighty-one members appointed by the Crown for life, though removable by the House itself for bankruptcy or crime; having co-ordinate powers of legislation with the House of Commons, except in the case of money or **tax** bills, which it can neither initiate nor amend; having no power to try impeachments; having the same privileges, immunities and powers **as** the English House of Commons when defined by law.

4. A House of Commons of **two hundred and thirteen** members, elected for **five** years on a **very** liberal franchise **in** electoral districts, fixed by law in each province; liable to be prorogued **or** dissolved at any time by the Governor-General on the advice of the Council; having alone, the right to initiate money **or** tax bills; having the same privileges, immunities and powers as the English House of Commons when defined by law.

5. A Dominion Judiciary, composed **of** a Supreme Court, of a chief justice and five puisne judges, acting as a Court of Appeal for all the Provincial Courts; subject to have its decisions reviewed on appeal by the Judicial Committee of the Queen's Privy Council **in** England; its judges being irremovable except for cause, on the address of the two Houses to the Governor-General.

---

who has supervision of all matters affecting administration of justice, and is legal adviser of the Dominion Government; Minister of Trade and Commerce, who has control of all matters relating to trade, and of customs and excise—the Controllers of which are now in the Cabinet; Minister of Finance, who has charge of finances and expenditures; Minister of Agriculture, who has also charge of public health and statistics; Minister of Marine and Fisheries, who has exclusive jurisdiction over sea coasts and inland waters; Minister of Interior, who has management of North-West lands, immigration, and Indians; Minister of Militia and Defence; Minister of Public Works; Minister of Railways and Canals; Postmaster-General—all of whose duties are obvious; Secretary of State, who is **also** Registrar-General, and has charge of correspondence, public printing, etc. In addition to these departmental heads there are always one or more Cabinet Ministers without portfolio. See Bourinot's "How Canada is Governed," 2nd ed., **pp.** 78-86.

The several authorities of government in the Provinces may be briefly described as follows:—

1. A lieutenant-governor appointed by the governor-general in council, practically for five years: removable by the same authority for cause; exercising all the responsibilities and powers of the head of an executive, under a system of parliamentary government; having no right to reprieve or pardon criminals.

2. An executive council in each province, composed of certain heads of departments, varying from five to twelve in number; called to office by the lieutenant-governor; having seats in either branch of the local legislature; holding their positions as long as they retain the confidence of a majority of the people's representatives; responsible for and directing legislation; conducting, generally, the administration of public affairs in accordance with the law and conventions of the constitution.*

3. A legislature composed of a legislative council and an assembly, in Nova Scotia and in Quebec; and of only an assembly or elected house in the other provinces. The legislative councillors are appointed for life, by the lieutenant-governor in council, and are removable for the same reasons as senators; cannot initiate money or tax bills, but, otherwise, have all powers of legislation; cannot sit as courts of impeachment. The legislative assemblies are elected for four years in all cases, except in Quebec, where the term is five; liable to be dissolved at any time by the lieutenant-governor, acting under the advice of his council; elected virtually on manhood suffrage in all the provinces except Nova Scotia and Quebec, where the franchise, however, is very liberal.

*The Executive Council, in all the provinces, comprises an Attorney-General, Commissioner or Minister of Mines and Crown Lands, Treasurer, Minister of Agriculture, Minister of Education (in the majority of cases), Secretary and Registrar, Commissioner of Public Works. The Prime Minister is generally the Attorney-General, but not necessarily so. Members of the council are also appointed without office. See Bourinot's "How Canada is Governed," pp. 148-152.

4. A judiciary in each of the provinces, appointed by the governor-general in council; removable only on the address of the two Houses of the Dominion Parliament.*

As regards the territories of the Northwest, they have been divided into five districts for purposes of government. Keewatin is under the control of the Government of Manitoba, but only until the question of boundaries will be finally settled. The other districts are governed by a lieutenant-governor and an assembly elected by the people in accordance with the statutes passed by the Dominion Parliament. The lieutenant-governor is appointed by the governor in council, and holds office on the same tenure as the same officials in the provinces; and while responsible government in the complete sense of the term does not yet exist in the territories, the lieutenant-governor has the assistance of an advisory council selected from the majority in the assembly. The Territories are represented both in the Senate and House of Commons of Canada.

Coming now to the distribution of powers between the Dominion and Provincial authorities, we find that they are enumerated in sections 91, 92, 93 and 95 of the fundamental law. The 91st section gives exclusive jurisdiction to the Parliament of the Dominion over all matters of a general or Dominion character, and section 92 sets forth the exclusive powers of the provin-

---

*The courts of Canada are numerous and follow the titles and procedure of those of England as far as possible, though in Lower Canada the existence of the Civil Code, based on the Roman law and the *Coutume de Paris*, has necessitated some differences. The courts range from those of the Court of Appeal or Supreme Court, with highest jurisdiction over civil and criminal matters, to the county, district, division, and magistrates' courts with limited powers. Appeals lie to the Supreme Court of Canada and the Judicial Committee of the Queen's Privy Council, with certain limitations made by law. See Bourinot's "How Canada is Governed," pp. 176 et seq.

cial organizations. The classes of subjects to which the exclusive authority of the Dominion Parliament extends are enumerated as follows in the Act :—

The public debt and property. The regulation of trade and commerce. The raising of money by any mode or system of taxation. The borrowing of money on the public credit. Postal service. Census and statistics. Militia, military, and naval service and defence. The fixing of and providing for the salaries and allowances of civil and other officers of the government of Canada. Beacons, buoys, lighthouses, and Sable Island. Navigation and shipping. Quarantine and the establishment and maintenance of marine hospitals. Sea-coast and inland fisheries. Ferries between a province and a British or foreign country, or between two provinces. Currency and coinage. Banking, incorporation of banks, and the issue of paper-money. Savings-banks. Weights and measures. Bills of exchange and promissory notes. Interest. Legal tender. Bankruptcy and insolvency. Patents of invention and discovery. Copyrights. Indians and lands reserved for Indians. Naturalization and aliens. Marriage and divorce. The criminal law, except the constitution of the courts of criminal jurisdiction, but including the procedure in criminal matters. The establishment, maintenance and management of penitentiaries ; and lastly, "such classes of subjects as are expressly excepted in the enumeration of the subjects assigned by this Act exclusively to the legislatures of the provinces."

On the other hand, the exclusive powers of the provincial legislatures extend to the following classes of subjects :—

"The amendment from time to time, notwithstanding anything in the Act, of the constitution of the province, except as regards the office of Lieutenant-Governor. Direct taxation within the province in order to the raising of a revenue for provincial purposes. The borrowing of money on the sole credit of the province.

12

The establishment and tenure of provincial offices and appointment and payment of provincial officers. The management and sale of the public lands belonging to the province, and of the timber and wood thereon. The establishment, maintenance, and management of public and reformatory prisons in and for the province. The establishment, maintenance, and management of hospitals, asylums, charities, and eleemosynary institutions in and for the provinces other than marine hospitals. Municipal institutions in the province. Shop, saloon, tavern, and auctioneer and other licences, in order to the raising of revenue for provincial, local, or municipal purposes. Local works and undertakings other than such as are of the following classes :—(a) Lines of steam or other ships, railways, canals, telegraphs, and other works and undertakings connecting the province with any other of the provinces ; (b) Lines of steamships between the province and any British or foreign country ; (c) Such works as, though wholly situate within the province, are before or after their execution declared by the Parliament of Canada to be for the general advantage of Canada or for the advantage of two or more of the provinces. The incorporation of companies with provincial objects. Solemnization of marriage in the province. Property and civil rights in the province. The administration of justice in the province, including the constitution, maintenance, and organization of provincial courts, both of civil and criminal jurisdiction, and including procedure in civil matters in those courts. The imposition of punishment by fine, penalty or imprisonment, for enforcing any law of the province made in relation to any matter coming within any of the classes of subjects above enumerated. Generally all matters of a merely local or private nature in the province."

Then in addition to the classes of subjects enumerated in the sections just cited, it is provided by section 93 that the legislatures of the provinces may exclusively legislate on the subject of education, subject only to the power of the Dominion Parliament to make remedial laws in case of the infringement of any legal rights enjoyed by any minority in any province—a provision intended to protect the separate schools of the Roman Catholics and the Protestants in certain provinces. The Dominion and the provinces may also concurrently make laws in relation to immigration and agriculture, provided that the Act of the province is not repugnant to any Act of the Dominion Parliament; and under section 94, the Dominion Parliament may provide for the uniformity of laws relative to property and civil rights in Ontario, Nova Scotia, and New Brunswick.

The statesmen that assembled at Quebec believed it was a defect in the American constitution to have made the National Government alone one of the enumerated powers, and to have left to the States all powers not expressly taken from them. For these reasons mainly the powers of both the Dominion and Provincial Governments are stated, as far as practicable, in express terms, with the view of preventing a conflict between them; the powers that are not within the defined jurisdiction of the Provincial Governments are reserved in general terms to the central authority. In other words, "the residuum of power is given to the central instead of to the provincial authorities." In the British North America Act, we find set forth in express words:

1. The powers vested in the Dominion Government alone.

2. The powers vested in the provinces alone.

3. The powers exercised by the Dominion Government and the provinces concurrently.

4. Powers given to the Dominion Government in general terms.

The effort was made in the case of the Canadian constitution to define more fully the limits of the authority of the Dominion and its political parts; but while

great care was evidently taken to prevent the dangerous assertion of provincial rights, it is clear that it has
the imperfections of all statutes, when it is attempted
to meet all emergencies. Happily, however, by means
of the courts of Canada, and the tribunal of last resort
in England, and the calm deliberation which the
Parliament is now learning to give to all questions of
dubious jurisdiction, the principles on which the federal
system should be worked are, year by year, better
understood, and the dangers of continuous conflict
lessened.

I have now reviewed the leading features of the
constitutional development of Canada, and shown in
what respects it is based on the American and British
systems of government. Englishmen generally will
assuredly find some satisfaction in the fact that their
greatest dependency has endeavoured to follow so closely
the leading principles of the parliamentary government
of the parent state. The constitution, on the whole,
appears to be a successful effort of statesmanship, and
well adapted to promote the unity of the Dominion, if
worked in a spirit of compromise and conciliation.
Canada is now governed by a political system which,
from the village or town council up to the Parliament
of the Dominion, is intended to give the people full
control over their own affairs. At the base of the
entire political organization lie the municipal institutions. Each province is divided into distinct municipal districts, whose purely local affairs are governed
by elected bodies, in accordance with a well-matured
system of law. Still higher up in the body politic is
the province with a government, whose functions and
responsibilities are limited by the federal constitution.
Then comes the general government to complete the
structure—to give unity and harmony to the whole.
With a federal system which gives due strength to the
central authority, and at the same time every freedom
to the provincial organizations ; with a judiciary free

from popular influences, and distinguished for character
and learning ; with a public service, resting on the safe
tenure of good behaviour ; with a people who respect
the laws—the Dominion of Canada must have a bright
career before her, if her political development continues to be promoted on the same wise principles that
so far illustrate her constitutional history.

CHAPTER III.—APPENDIX.

The English Privy Council and the Constitution of Canada.

BY A. H. F. LEFROY, M.A. (OXON.), BARRISTER-AT-LAW, TORONTO,
AUTHOR OF "THE LAW OF LEGISLATIVE POWER IN CANADA."

THE work that has been done during the last thirty years upon the Constitution of the Dominion of Canada by the Judicial Committee of the Privy Council in England, is well calculated to impress the imagination. The conventions of the Canadian Constitution—to adopt the distinction so clearly drawn by Professor Dicey—are borrowed from those of the Mother Country, but for its law, so far at least as concerns the formation and powers of the law-making bodies, we have to look primarily to the Imperial Statute passed in 1867, and known as the British North America Act. That may be termed the written portion of the Canadian Constitution, and there seems to be no reason why it should not prove as permanent as has been the Constitution of the United States. The latter, indeed, cannot be amended except by a concurrent vote of not less than sixty-six legislative chambers in the various States besides that of a two-thirds majority in each house of Congress, while the former might be amended or indeed repealed at any time by Act of the same Imperial Parliament which enacted it. But there is small likelihood of any such Imperial intervention, unless in compliance with resolutions passed in the federal and in all the provincial legislatures in the Dominion, at any rate so far as concerns the respective powers of those legis-

latures. In finally settling, therefore, as the Court of
last resort, the principles which should be applied to
the construction of the British North America Act, the
Judicial Committee are doing a very impressive work.

The Act itself does no more than mention in a general
way the lines of division by which legislative power
over the internal affairs of the Dominion is apportioned
between the federal and the provincial legislatures,
and no doubt it would have been unwise to attempt to
define and mark out with greater exactness the nature
of the laws which those bodies might respectively enact.
But the skeleton thus provided has to be clothed with
the flesh and blood of a living Constitution, and this
creative work has been placed upon the Judicial
Committee, assisted by the industry and learning
of the Canadian judges who preside over the Courts of
prior resort ; nor is the work a light or easy one. The
distribution of legislative power in the Act is comprised
mainly in two sections, one dealing with the jurisdiction
of the Dominion Parliament and the other with that of
the provincial legislatures, and both alike enumerate
certain large classes of matters the power to make laws
in relation to which, they provide, shall rest exclusively
with the former or with the latter. But it was impossible, as might have been expected, to avoid over-lap-
ping. Thus to the Dominion Parliament is given exclusive power to raise money by any mode or system of
taxation, but direct taxation within the province in
order to the raising of a revenue for provincial purposes belongs exclusively to the provinces ; the regulation of trade and commerce is exclusively in the
Dominion, but shop, saloon, tavern, auctioneer, and
other licenses in order to the raising of a revenue for
provincial, local or municipal purposes, is exclusively
in the provinces ; banking, bills of exchange and
promissory notes, interest, legal tender, bankruptcy
and insolvency, patents of invention and discovery,
and copyrights are exclusively assigned to the Domin-

ion, but property and civil rights in the province to
the provinces; marriage and divorce is exclusively for
the Dominion, but the solemnization of marriage in
the province exclusively for the provinces.

The task of reconciling and interpreting these provisions is, however, lightened by the enactment that the
enumerated powers of the Federal Parliament shall
belong to it "notwithstanding anything in this Act,"
and by a further provision which the Board have interpreted to mean that Parliament may deal with provincial matters wherever such legislation is necessarily
incidental to the exercise of its own enumerated powers;
and upon these clauses is based the predominance of
Parliament where *intra vires* legislation on its part
absolutely conflicts with provincial legislation which
would otherwise be of valid force.

But for all that it was of course soon discovered
that it could not have been intended that the powers
exclusively assigned to the provincial legislatures
should be absorbed in those given to the Dominion
Parliament, and therefore it has devolved upon the
Courts and ultimately of the Judicial Committee,
in the words of the latter, "however difficult it may
be, to ascertain in what degree, and to what extent,
authority to deal with matters falling within the
different classes of subjects exists in each legislature,
and to define in the particular case before them the
limits of their respective powers." And in this is
included the difficult and most important task of
defining and explaining the general residuary power
which is conferred upon the Federal Parliament to
make laws for the peace, order and good government
of the Dominion in matters not coming within the
classes of subjects assigned exclusively to the provincial
legislatures.

It would indeed be rash to predict that any portion
of the work thus laid upon the Privy Council will
prove unimportant in the future life of the Dominion,

but its importance is of course in some cases more immediately obvious than in others ; and how it is being performed must always have a general interest from the point of view of political science, which with the growth of population and political liberty cannot fail to be concerned more and more with the distribution of legislative power between a central national authority and local law-making bodies. A very brief glance at the subject, however, with a view of illustrating the broad general interest thus claimed for it, is all that is possible here.

In a case which came up some years ago, where the Dominion Parliament had passed a general prohibitory liquor law providing in a uniform manner for the prevention throughout the Dominion of the retail sale of intoxicating liquors, but to be brought into operation or not by a system of local option, the objection was raised that such legislation entirely wiped out the power of the provinces to raise a revenue from licenses. The Privy Council, though they did not concur with the view of the Supreme Court of Canada that the law came within the regulation of trade and commerce, held, nevertheless, that it was a valid one as coming within the general residuary power of Parliament above referred to, laying down the important principle that an Act of the Dominion Parliament is not affected in respect to its validity by the fact that it interferes with the object and operation of provincial Acts provided that it is not in itself legislation upon, or within one of the subjects assigned to the exclusive jurisdiction of the province, which they held the Act in question certainly was not. And conversely, a few years later, where a province had passed an Act taxing banking institutions in proportion to their capital, and it was objected that this was *ultra vires*, because banking and the incorporation of banks was a subject exclusively assigned to the Dominion, and that if such provincial legislation were permitted

banks might be taxed out of existence, their lordships
met the objection by formulating the principle that if
on a due construction of the British North America Act
a legislative power falls within the subjects assigned
to the provinces, it is not to be restricted or its existence denied because it may be abused or may limit the
range which otherwise would be open to the Dominion
Parliament.

And so throughout, in entire conformity with
the statement in the preamble of the Act that the
Constitution thereby provided for Canada was to be
similar in principle to that of the United Kingdom, the
Privy Council have laid it down that the intention of
the Imperial Parliament was to confer upon the
Canadian law-making bodies authority as plenary and
as ample within the limits prescribed, as it itself in
the plenitude of its power possessed and could bestow.
Thus their lordships, have refused to recognize as
applicable to Canada any principle analogous to that
by which a legal limitation to the power of the separate States of the Union is found by American judges
to exist at the point where the action of the State
legislatures comes into conflict with the powers of Congress, and have pointed to the federal veto over provincial legislation as forming a distinguishing feature
of the Canadian Constitution.

It has been, however, legislation in respect to the
liquor trade in the Dominion which has brought
before the Board the most searching constitutional questions for decision. The case already referred to was
followed shortly afterwards by one in which a provincial legislature had passed an Act restricting the hours
and places in which liquor might be sold by retail in
the municipalities, and in holding the Act *intra vires*
the Judicial Committee of the Privy Council astutely
laid down the principle, which they were unquestion-
ably the first to formulate, that subjects which in one
aspect and for one purpose fall within the jurisdiction

of the local bodies, may in another aspect and for another purpose fall within the jurisdiction of the Dominion Parliament, and such they held was the subject of legislative dealing with the liquor traffic. And this most interesting rule they had occasion to emphasize and illustrate in their judgment on a late Canadian appeal in May, 1896, in which questions were submitted to them intended to define what, if any, powers of absolutely prohibiting the sale, manufacture, and importation of intoxicating liquors were possessed by the provincial legislatures, a case which may perhaps prove to be the culminating point and conclusion of the long constitutional fight which has been carried on in the Dominion between the strong temperance party there existing and the liquor trade. Their lordships' decision has been that such restrictive legislation may have a purely local and provincial aspect restricting the consumption of liquor merely within the ambit of the province, and not affecting transactions in liquor between persons in one province and persons in other provinces, or in foreign countries, and as such it would fall within the power of the provincial legislatures, while, in a national and Canadian aspect, it nevertheless is open to the Dominion Parliament to enforce prohibition by a uniform Act extending throughout Canada, and the provisions of such a general law would, where in conflict, supersede provincial legislation.

In conclusion it may be mentioned that at the time of this going to press, an appeal is pending before the Judicial Committee upon a number of questions submitted by the Governor-General in Council, directed towards settling whether, on the proper interpretation of the British North America Act, the beds of the great rivers and lakes of Canada are vested in the Crown as represented by the Dominion Government, or in the Crown as represented by the Provincial Government,— or to use a shorter, but technically less accurate

. phrase, whether they belong to the Dominion or to the provinces. Upon the decision depends the ownership of the fisheries in such waters, from which no inconsiderable public revenue may be derived. Incidentally, too, the question, of some interest to lawyers, whether the rule that riparian ownership extends 'to the middle thread' applies in the case of these large bodies of water, and if not on what principle it is excluded, is likely to be decided.

Thus the small tribunal which meets in the Council Chamber at Whitehall, and forms the Court of last resort for the whole empire, is deciding from year to year, questions pregnant with importance to the future national life of the Dominion; and doing so in a manner which commands the respect of the Canadian people.

CHAPTER IV.

Administrative Departments of the Dominion Government.

THE division of function as between the Parliament of Canada and the legislative bodies of the provinces is determined by the British North America Act, 1867 (Vict. 30 and 31, c. 3), and the administrative departments are controlled in accordance with the provisions of this Act. The Parliament of Canada is empowered to deal with all matters not expressly assigned to the Provincial authorities. It is, however, explicitly charged with the control of the following :—

The Public Debt and Property ; The Regulation of Trade and Commerce ; The Postal Service ; Defence ; Navigation ; Quarantine ; Currency and Coinage ; Legislation regarding Weights and Measures, Interest, Bills of Exchange, Bankruptcy, Patents, Copyrights, Indian Affairs, Marriage and Divorce, Criminal Procedure, Penitentiaries ; Indirect Taxation.

The following departments are engaged in the administration of government :—

Justice ; Militia and Defence ; Secretary of State ; Public Printing and Stationery ; Interior ; Indian Affairs ; Finance and Treasury Board ; Customs ; Inland Revenue ; Post Office ; Agriculture ; Marine and Fisheries ; Public Works ; Railways and Canals ; Geological Survey ; Trade and Commerce.

The capital of the public debt of the Dominion in 1896-97 was $218,816,263, and the interest charge

$10,772,488.   One-half of the debt was issued at four per cent., about one-fifth at three and a half per cent., and about one-fifth at three per cent.   The cost of management of the public debt is $165,400 per annum.

The estimated expenditure for the year ending 30th June, 1897, was $41,647,921.   Of this the provinces received by way of subsidy $4,239,500.

CHAPTER IV.—SECTION 1.

Public Lands of Canada.

By A. M. Burgess, Commissioner of Lands for the Dominion.

SECTION 109 of the Act of Confederation provides that "all lands, minerals and royalties belonging to the several Provinces of Canada, Nova Scotia and New Brunswick, at the Union, and all sums then due or payable for such lands, shall belong to the several Provinces of Ontario, Quebec, Nova Scotia and New Brunswick in which the same are situate or arise," etc.; subject, however (section 117), to the right of Canada to assume any lands or public property required for fortifications or for the defence of the country, and subject also to the limitations of section 108, by the operation of which Ordnance Lands and lands and water powers connected with canals became the property of the Dominion. In 1870, the territorial rights of the Hudson's Bay Company in Rupert's Land were surrendered to the Crown in consideration of a payment of £300,000 sterling, certain tracts surrounding the Company's trading posts, and one-twentieth of all the land in the Fertile belt, which is defined as being bounded on the south by the International Boundary, on the west by the Rocky Mountains, on the north by the North Saskatchewan River, and on the east by Lake Winnipeg, Lake of the Woods, and the waters connecting the same.

By the terms of union between British Columbia and Canada, all the undisposed of lands in that province within twenty miles of the Canadian Pacific Railway were transferred to the Dominion, and under a later arrangement compensation was made for the lands in that belt not available for the purposes of the Dominion by a transfer of three million acres in the Peace River district.

Briefly stated, all the Crown Lands in the provinces which were prior to Confederation independent colonies—Prince Edward Island, Nova Scotia, New Brunswick, Quebec and Ontario, and in British Columbia except as to the Railway Belt and the tract of three million acres in the Peace River region, and the Ordnance and Canal Reserves—are vested in and administered by the several Provincial Governments. The Government of Canada owns and administers through the Department of the Interior the Ordnance and Canal properties in the older provinces, the Railway Belt and Peace River lands in British Columbia, and all the public lands in Manitoba and the North-West Territories, including the timber and minerals; but in British Columbia the precious metals, according to a decision of the Imperial Privy Council, are vested in the Crown in the right of the province, irrespective of whether the land itself is provincial or federal property.

Nearly two million acres of the public domain of Nova Scotia remain vested in the Crown, the small proportion of which fit for cultivation is situated remote from churches, schools and settlements, and is open for sale at twenty-five cents per acre for agricultural purposes. The minerals, however, are excluded from the grants in such cases, and when worked are subject to royalty rates. In New Brunswick almost eight million acres remain at the disposal of the Crown. The Labour Act practically provides for free grants of 160 acres each to actual settlers, upon condition that certain labour shall be performed on the public roads, residence upon the land for not less than three years in a house of given dimensions, and the cultivation of a minimum of ten acres. Areas of 200 acres may also be sold at public auction on settlement conditions at an upset price of a dollar and a quarter per acre. In Prince Edward Island there are practically no agricultural lands left vested in the Crown, and only a limited area of the property acquired by the Government of the province in 1875 from the landlords remains even under lease, nearly all the farm holdings being owned in fee simple by the occupants.

The area of Quebec could not be definitely ascertained until within the past twelve months, because of uncertainty as to its northern boundary, but the provincial authorities stated it in general terms as being 228,900 square miles. The true legal boundary of the province on the north would have been the dividing line between New France and the British possessions in North America, but that line had not been ascertained or agreed upon when New France was ceded to Great Britain, and subsequently portions of what was indisputably French territory were placed under the jurisdiction of Newfoundland. The adoption of the true legal boundary having thus been rendered impracticable, a conventional line, consisting practically of the East Main and Hamilton Rivers, connected by a due east and west line in latitude 52 degrees and 55 minutes, was agreed upon between the Governments of Canada and Quebec, and now only awaits legislative ratification. Under this arrangement the area of the province would be 347,000 square miles. The increase does not consist in any material degree of land valuable either for its agricultural, mineral or timber resources. Some six million acres of the ungranted Crown property of the province have of recent years been surveyed for settlement; and colonization—chiefly by repatriation from the United States—is being actively promoted by the Provincial Government, aided by the Dominion to some extent. Much of the northern part of the province has been only partially explored, and it is not possible to arrive at an intelligent estimate of the acreage fit for agricultural and pastoral purposes which is still at the disposal of the Government. In the Lake St. John, Gatineau, Temiscamingue and other districts, there are exten-

13

sive fertile tracts, surveyed and open to settlers at
prices varying from twenty to sixty cents an acre, subject to conditions of actual residence and cultivation.

The public lands in Manitoba and the North-West
are surveyed into townships of thirty-six square miles
described by consecutive numbers northward from the
International Boundary, which is the first base line,
and consecutive ranges westward from meridians four
degrees apart; and the townships in their turn are
sub-divided into sections of one square mile, each section containing about 640 acres. Township outlines
on the east and west sides run for a distance of four
townships on a due north and south course from one
correction line to another, or more correctly speaking
they run north and south from the base line to the
correction line a distance of two townships in each
direction. Owing to the convergence of these lines,
the result of the spherical form of the earth's surface,
the southern row of sections in each block of four
townships is somewhat smaller in area than the northern; but the deficiency or overplus, as the case may
be, being equally distributed among all the sections in
each row, the difference is practically unrecognisable.
In the country lying between the Lake of the Woods
and the Rocky Mountains on the east and west and
between the north branch of the Saskatchewan and
the International Boundary on the north and south—
the Fertile Belt of the Hudson's Bay Company—that
Company obtains its one-twentieth of the land by taking the whole of sections eight and twenty-six in each
township numbered by five, and the whole of section
eight and three-quarters of twenty-six in every other
township; while the Parliament of Canada has withdrawn from ordinary sale and from settlement sections
eleven and twenty-nine in every township throughout
the public lands under its control within the country
generally described as Manitoba and the North-West
Territories, to be sold by public auction only, after by

reason of settlement of the surrounding country they
have attained their fair maximum value, as an endow-
ment for the benefit of the public schools, the proceeds
of the sales being invested in Dominion securities, and
the interest paid over to the province or provinces
from time to time. Of the remaining sections, those
bearing even numbers are generally speaking open for
free settlement, 160 acres being granted to each settler
on condition of actual residence and cultivation for a
period of three years; and those bearing odd numbers
are utilized as subsidies to aid in the construction of
railways. The railway companies dispose of their
lands at reasonable prices, averaging about three dollars
per acre, for exceptional choice tracts the price being
of course higher; and the Government sells at the
same price to settlers desiring to enlarge their holdings
to a limited extent beyond 160 acres. The section of
Canada thus generally described is of enormous area
—about two and a half million square miles, in round
numbers—and of about two hundred and fifty million
acres fit for agricultural and pastoral purposes, only a
small proportion has yet been taken up by settlers, who
have hitherto confined their selections to the southern
half of the Province of Manitoba and the portions of
the Territories lying adjacent to railways in operation
and the vicinity of the rivers. The timber and the
minerals are reserved from ordinary agricultural grants,
and are administered according to the provisions of
special regulations made and promulgated by authority
of the Governor in Council. Settlers get all the timber
needed for ordinary building and fencing, and if it
cannot be found on their holdings, they are entitled to
free permits to cut what may be requisite on Government property. The whole public domain of Canada
is open to selection for mining purposes by persons
who may find minerals therein, and special privileges
are granted to original discoverers in such cases. An
office fee is payable for entering in the records of the

Department of the Interior and protecting for one year from the date of the entry any discovery of gold or silver, but the location has to be worked during the period in the manner and to the extent prescribed by the mining regulations and the entry renewed each twelvemonth. Quartz mining claims may be purchased outright on payment therefor at the rate of five dollars per acre after a prescribed amount has been expended in development. It is important to note that the coal deposits in the Territories are of enormous but so far unascertained extent, that they are being developed at several points hundreds of miles apart, and that the fuel supply of the future is beyond the region of speculation, independently altogether of the timber resources of the country.

British Columbia, covering an area of 350,000 square miles, has proportionately a small area of public land open for settlement, although in the aggregate there are thousands of square miles of arable soil, suited to the production of fruit and cereals and to the grazing of cattle, which are still vested in the Crown and open to free settlement on conditions of actual residence and cultivation. The prevalence of minerals, especially gold and silver, over the greater part of the main land, and especially in the southern section, gives much additional value to farming lands, the products of which can be disposed of in mining camps. The mines of that section have also furnished an accessible and profitable market for the cattle of the Alberta ranges. The agricultural lands in the railway belt are to be had by actual settlers for one dollar per acre, but the less conveniently situated lands of the province are mostly free. The settler gets twenty-five acres of timber with his holding in the belt, the heavily timbered tracts being held, as in the Eastern Provinces, for disposal by competition to the proprietors of saw-mills.

In the Yukon country there are no agricultural lands, and the supplies for the mining camps have all to be imported.

CHAPTER IV.—SECTION 2.

Lands Reserved for Parks.

By A. M. Burgess.

ROCKY MOUNTAINS PARK.

THE discovery in the vicinity of Banff Station, on the line of the Canadian Pacific Railway and east of the summit of the Rocky Mountains, of a series of hot springs, which upon analysis proved to be of great sanitary value, led to legislation by Parliament during the session of 1887 reserving a tract of land, designated as the Rocky Mountains Park of Canada, 260 square miles in extent. The Act sets out that the tract is reserved and set apart as a national park and sanatorium for the people of Canada, and authority is given to the Governor in Council to make regulations for its administration, and to appoint officers to give effect to such regulations. In addition to the attraction afforded by the mineral springs, the reservation embraces some of the most magnificent mountain, lake, and river scenery in the world, and liberal Parliamentary appropriations having been made for the purpose, these have all been rendered easily accessible to visitors by the construction of roads, bridges, and bridlepaths, under the supervision of a skilled superintending engineer. The waters of the hot springs have been made available for public use in various forms, and are found to be possessed of highly curative properties, especially in cases of rheumatic affections. Hotels and sanatariums have been erected at several points in the vicinity of the springs, the chief of which are a large, commo-

dious and handsomely fitted hotel belonging to the
Canadian Pacific Railway Company, and the Sana-
tarium of Dr. R. G. Brett, to both of which thousands
of tourists from all over the world resort every year.
These hotels are furnished with baths, to which water
direct from the hot springs is supplied by the Government. Special provision is made for the preservation
of game and forest trees, and heavy penalties imposed
for the infringement of the laws and regulations in
that behalf.

FOREST PARKS.

Section 78 of Chapter 54 of the Revised Statutes of
Canada authorizes the Governor-General in Council,
for the preservation of forest trees on the crests and
slopes of the Rocky Mountains and for the proper
maintenance throughout the year of the volume of
water in the rivers and streams which have their
sources in such mountains and traverse the North-West
Territories, to reserve unappropriated public land for
forest park purposes and to appoint necessary officers.
A fine is prescribed of not more than one hundred and
not less than ten dollars for cutting or injuring trees
in such forest parks, three months' imprisonment being
the penalty in default of payment. At several points
in the mountains the authority thus conferred has been
invoked and parks established, notably one lying adjacent to the Rocky Mountains Park and including
the two lakes known as Lake Louise and Lake Agnes,
situated in the heart of the mountains at considerable
altitudes, and of great natural beauty; and another at
Albert Canon on the Illecillewaet River in British
Columbia, a point of much interest to tourists.

CHAPTER IV.—SECTION 3.

Quarantine Service.

By F. Montizambert, M.D., F.R.C.S., D.C.L., General Superintendent of Canadian Quarantines.

THE whole of the Quarantine Service of Canada is under the administration of the Minister of Agriculture.

There is a medical officer as General Superintendent of Canadian Quarantines.

Each Quarantine Station is in the immediate charge of a specially appointed medical quarantine officer.

At each unorganized maritime or inland Quarantine Station the local Collector of Customs is the quarantine officer, with power to call in medical assistance when required.

The Quarantine Stations of Canada are:

1. On the Atlantic coast:

(a) Grosse Isle, in the River St. Lawrence, with Rimouski, the Louise Embankment and the Grand Trunk Wharf at Lévis, as sub-stations, Province of Quebec.

(b) Halifax, the harbour and Lawlor's Island, in the Province of Nova Scotia.

(c) St. John, the harbour and Partridge Island, in the Province of New Brunswick.

(d) Sydney, Cape Breton, in the Province of Nova Scotia.

(e) Pictou, in the Province of Nova Scotia.

(f) Hawkesbury, in the Province of Nova Scotia.

(g) Chatham, in the Province of New Brunswick.

(h) Charlottetown, in the Province of Prince Edward Island.

2. On the Pacific Coast :

(*a*) William Head, including Albert Head, in the Strait of Fuca, Province of British Columbia, and also including as a sub-station the port of Victoria ; and,

(*b*) Vancouver.

3. Every other port, on both coasts, each such port being designated an Unorganized Maritime Quarantine Station.

4. And every inland Customs port on the Canadian frontier, between the Pacific and Atlantic Oceans, each such port being designated an Unorganized Inland Quarantine Station.

Every vessel arriving from any port outside of Canada at any organized Quarantine Station is inspected by a duly appointed quarantine officer, at the place duly appointed for such inspection, and is not allowed to make customs entry at any port in Canada until it has received a clean bill of health.

No person is allowed to land from any vessel until such person has been declared by a quarantine officer free from infectious disease, and until, in the judgment of such officer, such landing can be affected without danger to the public health.

Coasting vessels from Newfoundland and from ports in the United States contiguous to Canada and free from infectious disease may, from time to time, be excepted from the quarantine regulations by order of the Minister of Agriculture.

Any of Her Majesty's ships of war, or any transport having the Queen's troops on board, accompanied by a medical officer and in a healthy state, is exempt from quarantine inspection and detention.

All quarantine inspections at the regular stations are made without any charge against the vessel.

All costs incurred in the maintenance of healthy persons, who may have been exposed to infection, detained for quarantine of observation, are at the charge of the vessel.

Persons actually sick are treated and taken care of in the quarantine hospital at the charge of the Government.

The appliances, materials and labour for disinfection are supplied by the Government without charge to the vessel.

The graver quarantinable diseases are: Asiatic-cholera, small-pox, typhus fever, yellow fever, and the plague. The minor: scarlet fever, enteric fever (typhoid), diphtheria, measles and chicken-pox.

In addition to the above recital, it is the duty of every quarantine officer to satisfy himself as to the presence or absence of any other contagious or infectious disease.

With respect to leprosy it is the duty of every quarantine officer, particularly on the Pacific coast, to satisfy himself as to the fact of the presence or absence of such disease among the passengers; and in the event of any case of such disease being found, the person affected is not allowed to land, but must be taken back by the vessel to the place whence he or she came.

Every passenger is required to furnish evidence, to the satisfaction of a quarantine officer, of having been vaccinated, or having had the small-pox.

During a time of cholera or other epidemic, the luggage of immigrants or passengers by every vessel arriving at any port in Canada, whether from an infected or healthy port or country, may by the direction of the Minister of Agriculture, be disinfected in each case.

Rags coming from a port or country in which infectious disease prevails may be prohibited.

New merchandise in general is accepted without question.

The disinfectants chiefly used are steam, superheated, under pressure and in a vacuum; solution of mercuric chloride; and sulphur dioxide. Formaldehyde is at present under trial, and will quite probably be adopted for quarantine use in the near future.

No quarantine officer nor other person employed in the Quarantine Service of Canada is permitted directly or indirectly to receive or take any fee or private gratuity or reward for any service rendered to any company, or owner, master or crew, passenger or other person at or detained in any quarantine, maritime or inland.

Breaches of the regulations are punishable by a fine of $400, and imprisonment for six months.

Every quarantine officer is empowered to give any necessary order, or do any necessary act, to enforce these regulations, and it is his duty to report immediately to the Minister of Agriculture any breach or attempted breach of them.

CHAPTER V.

SECTION 1.

Education in Canada.

By the Hon. G. W. Ross, F.R.S.C., LL.D., Minister of
Education of the Province of Ontario.

UNDER the distribution of legislative functions prescribed by the British North America Act, "Education" is assigned to the provincial legislatures, with the
reservation to the Dominion Parliament of the right to
intervene, within certain limitations, for the protection
of the rights of religious minorities. There is, therefore,
no "System of Education" in Canada as a whole,
though there is such a system in each province of the
Dominion, as there is also in that quasi-province
which is known as the "North-West Territories."

The pioneer provincial systems are those of Ontario
and Quebec, known prior to 1867 as Upper and Lower
Canada respectively; that of Quebec is quite unlike any
other educational system in Canada. Besides the "Department of Public Instruction," the head of which is
the "Superintendent," there is for administrative purposes a "Council of Public Instruction," divided into a
Roman Catholic Committee and a Protestant Committee. To the former is assigned the care of the Roman
Catholic schools and to the latter the care of the Protestant schools, making thus a complete cleavage
throughout the system. All the Bishops of the Roman
Catholic Church in the province are members of the

Roman Catholic Committee. The total number of
schools in Quebec at the date of the last report was
5,682, of which 4,727 were Catholic, and 955 were
Protestant. About five in every 1,000 pupils in the
Catholic schools were Protestants, and over eighty of
every 1,000 in the Protestant schools were Catholics.
More than 48,000 pupils of French origin were learning
English, and more than 18,000 of English origin were
learning French. The number of pupils studying
agriculture in the public schools was nearly 50,000.
Four schools in the province are specially devoted to
teaching agriculture, besides one dairy school. Of the
schools 5,004 were reported to be "elementary" and
678 "superior." The total average attendance of pupils
in the elementary schools of all kinds was 214,960 out
of a total of 284,047 enrolled. Though Quebec is
well-equipped with academical institutions, both Catholic and Protestant, it has none supported by the
state. Its universities are Laval (Roman Catholic),
in the city of Quebec; McGill (non-denominational),
in Montreal; and Bishop's (Anglican), in Lemonville.
in Montreal there is also a teaching branch of the
faculty of Laval.

The educational system of Ontario is under the administrative charge of a Minister of Education who
has an Educational Council to conduct all examinations
prescribed by the Department for the pupils of elementary and secondary schools. Roman Catholic
ratepayers have the privilege of establishing separate
elementary schools under certain limitations, but these
are all under the control of the Education Department
of which the Minister is the head. In Catholic districts the Protestant minority may establish separate
schools. The secondary school system is entirely non-
denominational and undivided. The total number of
public elementary schools in the province is 5,649,
taught by a total of 8,110 teachers of whom 2,662 are
men and 5,448 are women. The number of Catholic

separate schools is 328 and of Protestant separate
schools 10. The number of secondary public schools
is 129, taught by 554 teachers, and there are 90 kin-
dergartens taught by 184 teachers. The pupils enrolled in the elementary schools of all kinds number
493,063, with an average attendance of 272,211. There
are three classes of institutions for the professional
training of teachers in Ontario, namely (1) County
Model Schools, numbering 60 ; (2) two Normal Schools ;
and (3) one Normal College ; their non-professional
training is obtained in the secondary schools and universities. At Guelph, within fifty miles of Toronto,
there is an Agricultural College with a large farm attached, both maintained by the province, and at
Belleville and Brantford, respectively, there are institutes for the education of the dumb and the blind.
There are six universities in the province of which
only one, the University of Toronto, is a state institution. The others are Victoria (now "federated" with
Toronto); Queen's, of Kingston; Trinity, of Toronto;
McMaster, of Toronto ; Western, of London ; and the
University of Ottawa.

The educational systems of Nova Scotia, New
Brunswick, Prince Edward Island, Manitoba, British
Columbia and the North-West Territories closely resemble in their main features that of Ontario, with
two exceptions—the substitution of a Council and
Superintendent for the Minister of Education, and the
omission of Catholic separate schools. The Ontario
system had assumed something like a permanent form
before 1850, and at that time and down to 1876 it was
under the administration of a Chief Superintendent.
The provinces above named copied this feature of the
system, but they have not yet substituted for it a responsible Ministry of Education, as Ontario did over
twenty years ago.

When the Manitoba system was established in 1871,
it was given a dual form like that of Quebec, but in

1890 this was abolished, and the administrative machinery was consolidated under one Superintendent and an undivided Council of Public Instruction. The constitutional validity of the Act which made this change was contested in the Courts, but was ultimately affirmed by the Judicial Committee of the Imperial Privy Council. In order to meet the wishes of the Roman Catholic minority the school law of 1890 was amended in 1897 by the Manitoba legislature, but not so as either to re-establish separate schools or to re-divide the Council into denominational committees. In 1895 the average attendance of pupils was 19,516 and the number of teachers was 1,093, the sexes being about equally represented. There are three secondary schools in the province with an aggregate attendance of 385 pupils. There is only one university in Manitoba, and it is maintained by the province. It has not as yet any teaching faculty, but it examines and graduates students from several affiliated denominational teaching colleges.

In Nova Scotia at the date of the last report there were in operation 2,305 elementary schools, taught by 2,399 teachers of whom 540 were men and 1,859 women, and 18 county academies taught by 43 teachers. The total number of enrolled pupils in the elementary schools was 89,126, with an average attendance of 51,528 ; the attendance at the academies was 5,500. There are several universities, none of which are maintained by the state. King's College in Windsor, an Anglican institution, is the oldest chartered university in Canada. Acadia University in Wolfville, is maintained by the Baptists, and though Dalhousie University in Halifax is non-denominational it receives, the support of the Presbyterian church. St. Francois Xavier College and Ste. Anne's College, are under the auspices of the Roman Catholic church.

In New Brunswick there are 1,695 elementary schools under the charge of 1,790 teachers, and attended

by an average of 38,447 pupils. There are 13 County
Grammar Schools with an aggregate of nearly 3,000
pupils taught by 69 teachers. More than three female
teachers for one male teacher are employed throughout
the province. The University of Fredericton is maintained by the province. Mount Allison University
belongs to the Methodist church.

In Prince Edward Island the number of elementary
schools is about 450, and of teachers about 520. These
schools are attended on the average by nearly 14,000
pupils. Of the teachers about one-half belong to each
sex. For secondary education the province is dependent upon twenty High School departments attached
to elementary schools, and on one provincial High
School which draws its students from these departments. There is no university in the province.

British Columbia had over 200 schools in operation
in 1895, of which four were secondary. The teachers
numbered 319, and the average attendance was 8,610
out of an aggregate enrolment of 13,482.

CHAPTER V.—SECTION 2.

Provincial Public Buildings and Works.

By Kivas Tully, C.E., Public Works Department for Ontario.

AT the time of Confederation. 1st July 1867, the
only buildings which belonged to the Province
of Ontario were the Government House, and Old
Parliament and Departmental Buildings, the Asylum
for Insane, Osgoode Hall and Normal School, all
erected in the city of Toronto; the Reformatory for
Boys at Penetanguishene, the Court House and Gaol
at Sault Ste. Marie, and the Lunatic Asylums at
Orillia and Amherstburg.

Since Confederation the Province has erected a new
Government House, new Parliament and Departmental
Buildings in the Queen's Park, additions to the Lunatic
Asylum, Osgoode Hall and Normal School, Toronto,
and additions to the Court House and Gaol at Sault
Ste. Marie.

The Lunatic Asylum at Kingston was purchased
from the Dominion Government, and extensive additions have since been made to the buildings.

The Lunatic Asylum at Orillia was changed into an
Asylum for Idiots, and when abandoned a new Asylum
was erected on a different site.

New Asylums for the Insane have been erected at
London, Hamilton, Mimico and Brockville. Buildings
for the Deaf and Dumb and the Blind have been
erected at Belleville and Brantford respectively. A
Central Prison, a Reformatory for Females, and a
School of Practical Science, have been erected in
Toronto, an Agricultural College at Guelph, and

Dairy Schools at Strathroy and Kingston. Court Houses, Gaols and Lock-ups and Registry offices have been erected in the Muskoka, Algoma, Thunder Bay, Parry Sound, Nipissing and Rainy River Districts.

Large expenditures have also been made in the construction of locks, slides and dams, and improvements on inland waters.

The whole expenditure on Public Buildings and Works by the province since Confederation to date has been $10,500,000.

The Provinces of Quebec, Nova Scotia, New Brunswick, Prince Edward Island, Manitoba, and British Columbia have erected residences for the Lieutenant-Governors, and buildings for their Legislatures, besides other works and improvements not detailed in their reports.

The Dominion Government has erected a Government House, Court Houses and Gaols, Custom Houses, Immigrant Buildings, Offices, and a Lunatic Asylum in the North-West Territories.

In all the confederated Provinces the Dominion Government erects Post Offices, Custom Houses, Examining Warehouses, Quarantine and Immigration Buildings, Penitentiaries, Drill Sheds and Marine Hospitals.

The Governor-General's residence (Rideau Hall), and the Parliament Buildings at Ottawa were erected by the united Provinces of Upper and Lower Canada, but became the property of the Dominion under the British North America Act; large additions have been made to these buildings since the union.

Subsidies to railways have also been given by the Dominion and Provincial Governments, and large expenditures have been made on public works, for the construction of canals and other public improvements.

14

CHAPTER V.—SECTION 3.

The Financial Relations of the Provinces of the Dominion.

By the Hon. Richard Harcourt, M.A., Q.C., Provincial Treasurer for Ontario.

THE British North America Act of 1867, or Confederation Act, as it is commonly called, brought under one Federal Government the four Provinces of Ontario, Quebec, Nova Scotia and New Brunswick.

Under its provisions all the Customs and Excise duties, as well as certain other revenues of the provinces were transferred from the provinces to the Dominion. At the same time the Act in question by way of compensation provided that the following sums should be paid yearly to the several provinces for the support of their Governments and legislatures : Ontario, $80,000 ; Quebec, $70,000 ; Nova Scotia, $60,000 ; New Brunswick, $50,000 ; and also that an annual grant in aid of each province should be made equal to 80 cents per head of the population, as ascertained by the census of 1861, with a special provision in the case of Nova Scotia and New Brunswick.

This federal grant of 80 cents per head is generally spoken of as the provincial subsidy, and the grant for the purposes of government and legislation is called the specific grant.

In the case of the Province of Ontario, the two grants have amounted from year to year to $1,196,000. This subsidy and grant together, speaking generally, constitute the most important of the revenues of the provinces.

The Confederation Act enacted also that certain
assets which belonged at the date of the Union to the
old provinces of Canada should be the property of
Ontario and Quebec conjointly. It was also provided
in this Act that arbitrators should be appointed to
divide and adjust the debts, credits, liabilities, properties and assets of the old provinces of Upper and
Lower Canada. An arbitration was held and under
its findings certain special funds were allotted to
Ontario, and others to Quebec. The special or trust
funds thus allotted to Ontario aggregated $2,000,000,
and on this large capital sum it has received interest,
paid half yearly, ever since.

The Provinces of Ontario and Quebec have very valuable timber resources administered by their respective
Crown Lands Departments. The Crown Lands
Department of Ontario yields a large though variable
yearly revenue. The receipts vary with the lumber
market, and depend measurably on the amount of
timber cut and sold each year. The lessees of the
timber limits, as they are called, pay to the Crown in
the first place large bonuses for the privilege to cut
the timber, as well as dues on the amount of timber
they cut from year to year thereafter.

Since Confederation, the Province of Ontario has
received from its Crown Lands Department an average
yearly sum of about $900,000, and it is not feared that
there will be any serious diminution in this source of
revenue for very many years to come. This Province
has also been in receipt each year of large sums in the
way of interest on special funds which the Dominion
holds in trust for it, as well as on its own investments
—in municipal debentures for example. This interest
receipt has averaged since Confederation about $300,-
000 a year.

Both Ontario and Quebec have received large sums
yearly by way of fees paid by liquor licensees. This
receipt in recent years in Ontario has amounted to nearly

$300,000 annually. In Quebec a much larger sum has been received. Since 1892 considerable sums, amounting altogether to $648,000, have been received in Ontario as succession duties. This receipt is mainly a Collateral Inheritance Tax, and is set aside under the Act creating it for purposes of charity, such as maintenance of asylums and hospitals. Quebec also derives a considerable revenue from succession duties.

In Ontario sums ranging from $50,000 to $90,000 a year have been received from the sale of law stamps (paid by the litigants), and from several minor sources there is collected each year under the head of casual revenue, say $70,000. The development of the rich mines of gold, nickel and other ores of Ontario, which are now attracting the earnest and constant attention of foreign capitalists, will materially increase the provincial revenues. The receipts of 1897 from this source bid fair to be one of the striking features of the year's finances. These different sources provide practically all the revenue of the Province of Ontario, and the absence of other sources of revenue arising out of this or that mode of taxation commonly resorted to the world over is specially noticeable.

The average aggregate receipts of Ontario since Confederation have been about $3,000,000. In recent years, say during half of the intervening period, it has on the average exceeded that sum by about half a million of dollars.

The annual average expenditure of the Province of Ontario for the last thirty years has been about $3,000,000. For the last four years it has averaged more than three and a half millions of dollars. The largest single head of expenditure is that of maintenance of public institutions. Nearly $800,000 a year is spent for this purpose. The provincial asylums for the insane, the blind and the deaf and dumb are noticeable alike for the substantial character of the buildings themselves and for the skilful and humane treatment bestowed upon their inmates. The number of the afflicted classes thus cared for by the province exceeds 4,000. About three-quarters of a million of dollars is devoted annually by Ontario for educational purposes. Its schools, primary and secondary, and its colleges and universities, offer educational facilities unsurpassed in any country. The Public Schools, in which primary instruction is given to half a million of pupils, are open to all classes without payment of fees, and the cost of secondary education is noticeably low.

Ontario also spends about $200,000 a year for the promotion of Agriculture. It maintains at considerable expense one of the best Schools of Agriculture and Experimental Farms on the Continent.

A like amount is directly given by way of money grants in aid of hospitals and charities.

This province relieves the county municipalities in many ways, for example in the matter of the expenditures incident to the administration of justice, the prosecution of criminals, etc. Since Confederation it has paid on the average for this purpose about $230,-000 a year. During the last eight or ten years the average has been about $450,000 a year.

In addition to these large annual grants in aid of education, agriculture, asylums, hospitals, administration of justice, etc., the Province of Ontario has spent from time to time very large sums in the erection of public buildings and in aiding railways. The new Parliament Buildings alone, only recently completed, cost over $1,300,000. The asylum accommodation has at a great expense been doubled within the las few years. Since Confederation, nearly $8,000,000 have been spent on public buildings.

A similarly generous measure of aid has been given towards railway construction. The province has now 6,542 miles of completed railway in operation within its boundaries, and nearly $7,000,000 have been thus far actually expended in railway grants.

The Province of Ontario has practically no obligations, present or future, outside of its railway obligations (certificates issued to railway corporations to aid in their construction), which are distributed over a long period of years. Even taking these obligations into account, it has a surplus in securities, trust funds, and other assets, of several millions of dollars. It is especially rich in natural resources, actual and potential, and present indications point in the near future to a degree of prosperity far exceeding even that of the past.

CHAPTER V.—SECTION 4.

Unalienated Lands of the Public Domain of Ontario Suitable for Settlement.

By Thomas W. Gibson, Secretary of the Bureau of Mines,
Toronto.

THE area of the Province of Ontario has been vari-
ously estimated, but may be set down as about
200,000 square miles, or 128,000,000 acres. Of the
total area, 22,000,000 acres, as nearly as can be calculated, have passed into private hands for purposes of
settlement and mining—mainly the former—leaving
about 106,000,000 acres still the property of the
Crown. Roughly speaking, all that part of the province south of a line drawn from Penetanguishene on
the Georgian Bay, east to Arnprior on the Ottawa
River, is fertile land, eminently suited for farming,
and has long been the home of a prosperous agriculture. In this section are the cities and towns, the mills
and manufactories, the schools, colleges and churches,
as well as the great bulk of the population. Stretching north to James Bay and the Albany River, and
westward along the northern shores of Lakes Huron
and Superior and on to Lake of the Woods, lie the
Crown lands of Ontario—a vast expanse of territory
comprising a great deal of rough and broken land un-
inviting to the agriculturist, but containing also many
stretches of arable and fertile soil. The settled land
to the south for the most part overlies geological
formations belonging to the Devonian series, and is
composed of drift transported by the glaciers in the
long-past ice ages from the rocky uplands of the north.
The latter on the other hand consist of Laurentian

and Huronian rocks, with their various subdivisions, except on the extreme north, where the slope which drains into James Bay is partly underlaid by Devonian formations resembling those of southern Ontario. The older rocks are usually clothed with a scantier covering of soil than the newer series, but there are clay plains and fertile river valleys which rival in productive power anything found in the south. The height of land which divides the waters running into the great lakes from those running into James Bay, cuts off a territory containing perhaps 80,000 square miles, the greater part of which is not only at present inaccessible for lack of means of communication, but is almost wholly unexplored. To be available for settlement, lands must afford at least passable means of ingress and egress, and must not be too far from markets for their products. In this respect the wild lands of Ontario are well served. The Canadian Pacific Railway runs for nearly 1,000 miles through their full width from east to west; the great lakes with their numerous harbors lie on their southern border; and many navigable streams and inland lakes intersect and connect them in all directions, thus making access and transportation comparatively easy and cheap.

The principal tracts of land available for immediate settlement are as follows:

1. In the districts of Muskoka, Parry Sound, Haliburton and Nipissing, and parts of the counties of Peterborough, Hastings, Frontenac, Addington and Renfrew. Between five and six millions of acres in these districts have been opened for settlement, mainly under the provisions of the Free Grants and Homesteads Act, which give an actual settler one hundred acres if he be a single man, or two hundred acres if the head of a family, free of cost, on his performing settlement duties. These consist of five years' residence, clearing and cultivating at least fifteen acres, and erecting a habitable house. Much of these sections is rocky and

not adapted for tillage, but there are many arable
areas, and the running streams and nutritious grasses
which abound give ample scope for dairying and stock
raising.

2. At various points along the main line of the Canadian Pacific Railway, in the districts of Nipissing and
Algoma, such as Mattawa, North Bay, Sturgeon Falls,
Sudbury and Sault Ste. Marie ; also on St. Joseph and
Manitoulin Islands (the latter administered as Indian
lands by the Dominion Government). About 900,000
acres contiguous to these several places are offered to
settlers, partly as free grants, but mostly for sale at
twenty or fifty cents per acre. Three or four years'
residence, the erection of a habitable house, and clearing
and cultivating ten per cent. of the area, are required
previous to issue of Crown patent. Broken land is
here also the rule, though there are many tracts well
suited for farming, and some well-settled and prosperous neighborhoods.

3. At Port Arthur, where there are several Free
Grant townships, and at Dryden, on the C. P. R., between Port Arthur and Rat Portage, where two townships have been opened for sale, and two others recently surveyed and laid out. There is a large area of
strong clay land here equal in extent to a good-sized
county, easy to clear and free from boulders. A pioneer farm started by the Department of Agriculture in
1894 has led to a considerable settlement and a large
amount of land being taken up. The price is fifty
cents per acre, and the usual settlement duties are
required.

4. In the valley of the Rainy River, in the north-
western part of the province, the Ontario side of which
comprises a fertile alluvial belt of 500,000 or 600,000
acres. This section is one of much promise, as the
gold-mining industry of the Rainy Lake and Seine
River regions, now being rapidly developed, will afford
an excellent market for agricultural products. There

are twenty townships open for settlement under the Free Grants Act. Continuous steamer navigation connects Fort Frances at the foot of Rainy Lake, with Rat Portage on Lake of the Woods.

5. On Lake Temiscaming, where at least 500,000 acres of rich calcareous subsoil covered by many inches of vegetable mould is found in one block stretching to the north and north-west of the lake. Twenty-five townships have been surveyed, of which five have been placed on the market at fifty cents per acre to actual settlers. The timber is mostly small, and the land is not difficult to clear. Considerable settlement is now going in.

Besides these areas now available, there are many others of greater or less extent, which are yet inaccessible, or too remote from markets to be of immediate use. One of these is a tract of land containing many thousands of acres, well adapted for settlement, discovered in 1896 near the upper waters of the Montreal River, by an exploring party sent out by the Department of Crown Lands. In the valleys of the great rivers running northward into James Bay, or on the tablelands between. there is reason to believe large areas of good agricultural land exist.

There are no prairies in Ontario, and by far the greater portion of the tracts mentioned above is covered with timber of various kinds, such as pine, hemlock, cedar, spruce, tamarac, maple, elm, basswood, birch, etc. The pine is mostly reserved to the Crown on lands sold or given away for settlement, and usually belongs to the lumberman holding a timber license from the Government. The settler is allowed sufficient pine for building, fencing and fuel, and all other kinds of timber become his property on his purchasing or locating the land from the Crown.

As a rule, the cultivable land is good and well-watered, and wherever settlement has yet taken place in Ontario all the principal grains and products of the temperate zone ripen and do well, such as wheat, oats, peas, barley, potatoes, vegetables and root crops of every kind. Hay, both timothy and clover, is grown in abundance and with little effort. In some parts, indeed, it is frequently found growing wild from seed scattered by chance along the roads in the lumber districts. Horses and all kinds of farm stock thrive.

The white pine forests of Ontario are very extensive, and support a large and important lumbering industry. Wherever lumbering is carried on, a ready market is provided for agricultural products, particularly for hay, oats and beef, in the lumber "shanties," and prices are usually higher than those current in the older portions of the province. The present activity in mining affairs promises to add a large population engaged in opening up the mines, which will afford an equally good market for farmers' produce. There are numerous water powers on the rivers of northern Ontario, in proximity to a wealth of raw material in the shape of wood and timber. These can hardly fail to be largely utilized in the future as the seat of manufactures of many kinds. Saw mills, sash and door factories, pulp mills, match factories and other wood-working industries would here find two great requisites of success—abundance of raw material and cheap power.

The climate of northern Ontario does not greatly differ from that of other portions of the province, save that it includes a longer and somewhat colder winter. The altitude is not materially greater than in the south, there being the same absence of mountains or high hills as in all other parts of the country. The air is dry, and the winter days though cold are far from unpleasant. In summer the weather is warm, and, as a rule, there is a large proportion of sunshine. Summer frosts are not unknown, but droughts seldom or never occur and there is never an entire failure of crops from any cause.

Game, such as deer, partridges, etc., is abundant in

220 UNALIENATED LANDS OF THE PUBLIC DOMAIN.

most of the wooded districts, and along with the salmon trout, whitefish, herring and bass, so common in the lakes and streams of the province, often forms an agreeable change in the settler's diet.

Settlement in the new districts of Ontario has come mainly from the older parts of the province, and from among farmers or their sons, whose means, though insufficient to purchase the higher-priced lands of their own localities, were yet equal to giving them a start on a bush farm. Such men, accustomed to the axe and the plough, make the very best of settlers; but there are also many who have come direct from the old land and who after gaining the experience necessary for meeting the exigencies of bush life have prospered in their new surroundings. Persons of this class, however, would do better to purchase farms already wholly or partly cleared, as they would thus escape the hardships incidental to taking off the timber, and would begin life anew under conditions resembling those with which they are familiar. In a new and thriving settlement, schools and churches soon spring up, roads and bridges are built, usually with government assistance, townships are incorporated with powers of self-government, and the forest gives way to the farm. In all the frontier parts of Ontario the same process is going on as that which has made it one of the most advanced and prosperous communities in America. Ontario has the land; all it needs is the people to possess and occupy it.

CHAPTER V.--SECTION 5.

Provincial Lands Reserved for Parks and Game Purposes.

By Thomas W. Gibson, Secretary of the Bureau of Mines,
Toronto.

THESE comprise three areas or tracts of land varying widely in their character.

1. THE ALGONQUIN NATIONAL PARK OF ONTARIO.

The advisability of preserving a number of important rivers in full and equal flow, of securing the
beneficial effects upon climate which large and unbroken forests are generally conceded to produce, of
affording on a grand scale a health resort and recreation
ground which should be common to all the people, of
providing an asylum where game and furbearing
animals might breed and replenish their numbers in
peace and safety, and of undertaking experiments in
practical forestry, led the Government of Ontario in
1893, to set apart a tract of wild land in the District of
Nipissing containing, with subsequent additions, about
1,110,000 acres of land and water under the name of
The Algonquin National Park of Ontario, to achieve,
if possible, these important ends. The Park occupies
a tract of rocky country south of Lake Nipissing and
between the Ottawa River and Georgian Bay. It com-
prehends within its limits the watershed dividing the
rivers flowing east into the Ottawa from those running
north into Lake Nipissing and west into Georgian Bay ;
among these streams being the Amable du Fond,
Petawawa, Madawaska, South and Muskoka. Rivers,

lakes and streams so intersect the territory that a
traveller may go by canoe in almost any direction from
one side of the Park to the other. The land is for the
most part of no agricultural value, and is covered with
timber. It was originally one of the most extensive
pineries in the Province. Besides white and red pine,
hemlock, tamarac, spruce, cedar, maple, birch, ironwood
and basswood make up the bulk of the forest. A great
deal of the pine has been cut and removed, and large
quantities are still taken out every winter. There is,
however, sufficient of the other kinds of timber to
maintain the wooded character of the Park. Considerable areas have been burned over, which are now
covered with a second growth of cherry, poplar,
etc., and are the favourite feeding grounds of the
larger animals in the Park. The chief game animals
are the moose, the largest native of Canadian woods,
and the red deer. The furbearing animals are the
beaver, otter, marten, mink, fisher and muskrat. Of
these, the moose and beaver have in the past been
pursued so mercilessly as to be in danger of becoming
extinct throughout the whole of the Province. Hunting
and trapping in the Park are absolutely forbidden, and
the result has been a large increase in the number of
all kinds of animals, especially the beaver, which breed
very rapidly. They are now spreading beyond the
limits of the Park, where they may be taken in their
proper season, though moose and beaver are both protected until 1900. Of beasts of prey, bears and wolves
are numerous. The latter are very destructive to the
deer. The lakes teem with all the varieties of trout,
and many contain also whitefish and herring. Partridges are very plentiful, and wild ducks are also
found, though they are not numerous. The natural
beauties of the Park, and its attractions as a place of
summer resort, are very great. The hundreds of lakes,
large and small, the winding streams, the excellent
fishing, the gloom and glory of the primitive forest,

are all alluring to the holiday seeker and to the mind wearied with toil and care. The Ottawa, Arnprior and Parry Sound Railway runs through the south-western corner of the Park, and makes it accessible by rail. The affairs of the Park are managed by a superintendent and staff of rangers.

2. RONDEAU PROVINCIAL PARK.

A sandy peninsula runs south into Lake Erie from the county of Kent, and curving towards the west encloses Rondeau Harbor, a safe shelter for vessels exposed to the storms which sometimes come up suddenly on Lake Erie. It contains about 5,000 acres of Crown land, and is largely covered with a luxuriant growth of timber, comprising pine, hickory, oak, whitewood, black walnut, sycamore and soft maple, several of the varieties being unknown except in this district of the Province. Rondeau Harbor is the resort every autumn of great flocks of wild ducks which pause here in their southward flight to feed on the water celery and wild rice which grow in profusion along its shallow shores. Wild turkeys, formerly found in great numbers here, are being again introduced, as well as English and Mongolian pheasants and other game birds. A number of summer cottages have been erected on the Park, and it is fast becoming the favorite picnic and camping ground of this part of the Province.

3. QUEEN VICTORIA NIAGARA FALLS PARK.

The impositions and annoyances to which visitors to the cataract of Niagara had long been subjected grew to such a height that in 1885 steps were taken by the State of New York and the Province of Ontario to constitute the land immediately surrounding the Falls a public park, and to throw it open on both sides of the river for the use and enjoyment of all who chose to resort thither. On the Ontario side the territory included in the Park proper comprises 154 acres, but

with the chain reserve along the bank of the river from
Fort Erie to Niagara-on-the-lake and the lands below
the cliff, etc., the whole area is brought up to 675 acres.
This is vested in a board of commissioners for the
management of the Park appointed by the Lieutenant-
Governor in Council, whose administration of affairs so
far has not cost the Province a dollar.   A revenue sufficient to bear all charges of maintenance, improvements,
etc., is derived from rentals paid for the right to use the
water power to generate electricity, for the right to run
an electric railway along the bank from Queenston, and
similar sources.   The object aimed at has been fully
achieved, and travellers who come from all parts of the
world to see Niagara may now do so in peace and quiet-
ness, amid far more picturesque conditions than formerly, and free from the necessity of either submitting
to extortion or foregoing the pleasure of viewing the
rapids and falls from some of the most advantageous
spots on the Canadian side, which admittedly furnishes
a better sight of them than can be had from American
soil.

4.   QUEBEC.

By an Order in Council dated the 6th November,
1894, subsequently confirmed by the Act 58 Victoria,
Chapter 22, the Government of the Province of Quebec
set aside as a forest reservation, fish and game preserve
and public park and pleasure ground, a tract 2,531
square miles in extent, including the head waters of
the rivers Montmorency, Jacques Cartier, Ste. Anne de
la Parade, Batiscan, Metabetchouan, Upikauba, Upica,
Chicoutimi, Boisvert, A Mars, Ha! Ha!, Murray and
Ste. Anne's.   The statute empowers the Lieutenant-
Governor in Council to annex to this Park any adjoining unconceded Crown Lands.   Fishing, shooting, cutting timber, mining, and the sale of intoxicating liquors
within the park are all prohibited, and due provision
is made for the enforcement of the law.

CHAPTER VI.

Municipal Institutions in Canada.

By C. R. W. Biggar, M.A., Q.C., formerly Solicitor and
Counsel for the City of Toronto.

THE Imperial Statute 30-31 Vict. chap. 3, known as
"The British North America Act, 1867,"—which
is in effect, the Constitution of the Dominion of
Canada—assigns (section 92) to the Legislatures of the
various Provinces, exclusive power to make laws in
relation to certain enumerated classes of subjects,
including (No. 8), "Municipal Institutions in the
Province."

The object of this paper is briefly to outline the
nature of these municipal institutions in each of the
several Provinces.

It is interesting to trace the successive modes in which
our local affairs have been managed:—at first, through
the Governor or Lieutenant-Governor in Council:
next, through the agency of Courts of General Quarter
Sessions, consisting of justices of the peace appointed
by royal commission for each district in the Province,
and finally, by the rate-payers themselves, acting
through their district, county, and local councils.
The result (briefly stated) is, that at present every
inhabitant of the settled districts of Canada, except in
the Province of Prince Edward Island, is a member of
at least one of the various municipal corporations in
which is vested the power of taxation for local purposes, the control and management of the public high-

15

ways and public buildings of the municipality, and which manages these and similar local affairs through the agency of a municipal council, chosen, (in most cases annually), by the resident ratepayers of the locality.

Since it is in the Province of Ontario that this system is most highly developed and completely organized, it will be convenient first to consider somewhat in detail the provisions of the municipal law of that Province ; afterwards, and more briefly, the municipal institutions of the other Provinces as compared therewith.

ONTARIO.

The municipal divisions of Ontario are :—

1st. The minor or local municipal corporations, consisting of—

(A) *Townships*,—which are rural districts having an area of eight or ten square miles with an average population of about 3,000 souls.

(B) *Villages*,—with a population of not less than 750 persons within an area of 500 acres.

(C) *Towns*,—having a population of at least 2,000.

Municipalities of these three classes, (except such larger towns as have separated themselves for municipal purposes from their original share in the government of the county in which they are locally situated,) are grouped together into a larger organization, viz.

2nd. *County* municipalities, the affairs of which are under the control of a County Council.

The 3rd class of municipal corporations consists of :—

Cities,—which may be erected out of towns when the population of the latter exceeds 15,000. Their municipal jurisdiction combines that of both the local and the county municipality.

The government of each local municipality is vested in a council, annually elected by the adult ratepayers thereof, being men, unmarried women or widows, and subjects of Her Majesty by birth or naturalization, who are—

(A) Freeholders or tenants of property rated at an actual value in townships and villages of $100, in towns of $200 to $300, and in cities of $400, or

(B) Ratepayers upon an annual income of $400, or

(C) Farmers' sons, resident upon the farm of their father or mother for a twelvemonth prior to the Election.

A municipal councillor, reeve, or mayor must be a resident within the municipality or within two miles thereof, a subject of Her Majesty, a male, of the full age of twenty one years, not disqualified under the Act, and must hold, as proprietor or tenant, in his own name, or in the name of his wife, freehold or leasehold property, rated on the last revised assessment roll of the municipality, to at least the value following over and above all charges and incumbrances thereon :

In townships, freehold to $400 or leasehold to $800 ; in villages, freehold to $200, or leasehold to $400 ; in towns, freehold, to $600 or leasehold to $1,200 ; in cities freehold to $1,000 or leasehold to $2,000. The qualification of a county councillor is the same as that of councillors in towns.

The list of persons disqualified for municipal office in Ontario (and it is practically the same in the other English-speaking provinces) includes Judges and other officers connected with the administration of justice, registrars, municipal officials, persons licensed to sell spirituous liquors, license commissioners, issuers or inspectors of licenses, high-school trustees and persons working or interested in contracts with the corporation.

The list of those exempt includes members of Federal or Provincial Parliaments, civil servants, officers of the Courts of Justice, coroners, clergymen and ministers of every religious denomination, barristers and solicitors in actual practice and members of the medical profession, persons engaged in teaching, millers, firemen and all persons more than sixty years of age.

The council of each township or village consists of

a reeve, elected by general vote, and four councillors, one of the latter, for each 500 voters in the municipality, being styled " deputy reeve."

Town councils consist of a mayor, elected by general vote, and three councillors, chosen by and from each ward if the number of wards is four or less, or two councillors for each ward if there are five or more wards.

City councils are composed of a mayor elected by general vote, and three councillors elected by and from each ward.

For county council elections, each of the forty-two municipal counties of the province is divided into " county council divisions," which in the same county are as nearly as possible equal in population, but (according to the size of the county) may each contain from 2,500 to 7,500 persons. The electors of each county council division who are entitled to vote therein for local councillors, also elect biennially two members of the council of the county or united counties ; and the county council at their first meeting in each year elect from among themselves a presiding officer styled the " warden."

The meetings for the nominations of county councillors are held biennially in each county on Monday in the week preceding the first Monday in January ; those for the nomination of local councillors annually on the last Monday in December, the poll both for local and county councillors being held on the first Monday in January. The voting is by ballot upon voters lists prepared by the local municipal clerks from the last assessment roll of the municipality which has been revised as hereinafter mentioned, and every municipality is divided for polling purposes into sub-divisions, each containing not more than 200 electors. For each polling sub-division the local municipal council appoints a deputy returning officer, who presides over the poll held in his sub-division, and, at its close, counts

the ballots in the presence of the agents for the several
candidates, and certifies the result to the returning
officer, who in elections for local councils is the clerk
of the local municipality, and in county council elections the county clerk. The polling is conducted in
the same manner as at a parliamentary election, and
secrecy of voting is secured by very stringent enactments. The polls are open from 9 a.m. till 5 p.m.; and
in elections for local municipal councils, the result, as
ascertained from the certificates of the deputy returning officers, is announced by the clerk of the local
municipality at the town hall at noon on the following
day; in county council elections, by the county clerk
at the shire hall on the Monday following the election.

Provision is made for a recount of votes by the
County Judge, and for inquiry into the validity of
the election and the removal of any person illegally
elected by summary proceedings in the nature of a
quo warranto, which may be had either before a Judge
of the High Court, the Master in Chambers or a County
Court Judge.

Fraudulent voting is punishable by imprisonment,
bribery or undue influence at a municipal election, or
at the voting upon a by-law, by fine, disqualification
to vote, and (in the case of a candidate) ineligibility
for municipal office for two years thereafter.

The officers of every municipal council are:

1st. The head—mayor, reeve, or warden, as the case
may be—annually elected by the ratepayers in the
first two cases, and in the third, annually chosen from
among themselves by the county council. His duty is
simply to preside at council meetings and to oversee
the other municipal officers. He has no power to veto
or to refer back any by-law, resolution or other act of
the council, and no casting vote in the event of a tie.

2nd. The clerk—who keeps the records of council
meetings, and the documents belonging to the municipality.

3rd. The treasurer—both of these are practically permanent officers, though they really hold office during the pleasure of the council. The remaining officers (except the assessment commissioners of cities), are annually appointed.

4th. Auditors of municipal accounts.

5th. In county municipalities, the valuators, and in local municipalities, the assessors, who prepare annual lists of the rateable property within the municipality. In cities and towns there may be constituted a board of assessors under a presiding officer, styled an "assessment commissioner."

6th. (In local municipalities only), collectors of the annual rates imposed by the local and the county councils, as well as of those imposed for "local improvements," and for school purposes. There are no parliamentary rates or taxes.

The jurisdiction of each municipal council (except certain special powers as to extra-territorial expropriation) is confined to the municipality, and its legislative powers, like those of other corporations, are exercised by by-laws which are enforceable by summary proceedings before a justice of the peace ; convictions for a breach of any by-law being punishable by a fine not exceeding $50, collectable by distress, or, in default of payment, by imprisonment for twenty-one days, (in the cases of convictions for keeping a house of ill-fame in a city, six months), unless the fine and costs are sooner paid. Each council is required to make annual estimates of the sums necessary to be expended during the year for all purposes within its jurisdiction, and to assess and levy upon all rateable property within its jurisdiction, an annual rate sufficient to discharge all debts falling due within the year. The aggregate rate for all purposes, exclusive of school of taxes, and rates for improvements payable by local assessment upon the properties specially benefited thereby, is limited to two per cent. per annum upon the actual value of the rateable property within the municipality.

The voting upon a by-law creating a debt, is by ballot, is conducted in the same manner, and is subject to the same statutory provisions as to secrecy of proceedings, bribery, treating, undue influence, etc., as the polling at a municipal election. Provision is also made for a recount or scrutiny of the ballots by the County Judge upon the petition of any elector.

If money has to borrowed, *e.g.*, to defray the cost of permanent improvements or otherwise, the by-law authorizing the creation of the debt must first be approved by a majority of the rate-paying freeholders or leaseholders entitled to vote at municipal elections. The debentures issued for any debt so created, must be payable within thirty years at farthest, and the by-law must impose an annual rate sufficient to pay off the debt by annual instalments, or to pay the interest annually, and provide a sinking fund, sufficient, with interest at five per cent. per annum upon the investments thereof, to pay off the principal at maturity. If the debt is created for the purpose of a loan to, or for the purchase of stock in, for or on guaranteeing the repayment of money borrowed by a railway company, or for aiding by way of bonus a railway, or water-works, or water company, the by-law must receive the assent of one-third of all the ratepayers entitled to vote thereon, as well as of a majority of those actually voting.

Very full provisions are contained in the statute respecting the securities in which sinking fund accumulations may be invested; and if moneys raised for a special purpose are diverted therefrom, every member of the council who votes for such diversion, becomes personally liable for the amount so diverted, and is disqualified from holding any municipal office. In addition to the regular annual expenditure of the municipality (to be provided for by general rate), special power is given in regard to incurring debts for local improvements, the cost of which may be defrayed by special local rates, assessed and levied upon the property immediately benefited thereby. Such "local improvements" may be undertaken.

1. Upon petition of the ratepayers upon the street or place or within the area where the improvement is to be made, or

2. By proceedings initiated by the council of a municipality after notice to owners of the properties affected and intended to be assessed therefor, who may, by petition, prevent the council from undertaking the work or improvement.

It would be impossible here to enumerate in detail the subjects in respect of which municipal councils are authorized to legislate. Briefly summarized, they include :—

1. The organization and management of the municipality itself and the conduct of its public business.

2. The conduct of municipal elections.

3. The appointment, control and payment of municipal officers.

4. Provision for the payment of members of township and county councils for attendance at meetings of the council and for necessary travelling expenses, and of aldermen, chairmen of standing committees and members of the local Board of Health and Court of Revision in cities of over 100,000 inhabitants, (e.g., Toronto) and of the mayor, warden or other head of either a local or county municipality.

5. The finances of the municipality.

6. The protection of life and property therein, by establishing a police force, by regulating the erection of buildings, hoists and elevators, the size and strength of walls, fire escapes, fire engines, fire companies, dangerous traffic on streets, the running at large of dogs, the fencing of vacant and other property, etc., etc.

7. Public morals, by the suppression of disorderly houses, houses of ill-fame, gambling, horse-racing, drunkenness, profanity, indecent exposure, etc.

8. Public health, including power to abate public nuisances, to regulate noxious trades, to prohibit the sale of articles unfit for food, to regulate the occupation of dwellings in crowded or insanitary localities, to provide pure water and compel its use within the municipality, to compel proper sewerage, drainage and other sanitary measures.

9. The control of the public highways and bridges, wharves and harbours.

10. The beautifying of the municipality by encouraging the planting and care of trees.

11. The establishment and control of fairs and markets.

12. The licensing and regulation of various trades within the municipality, e.g., those of auctioneers, billposters, hawkers and pedlars, milk-vendors, plumbers, tobacconists, and keepers of bowling alleys, billiard tables, exhibitions and shows, intelligence offices, victualling houses, etc.

13. Aiding general and technical education by the erection of school houses, the establishment of prizes, scholarships and other rewards, the establishment and maintenance of schools for artisans, etc.

14. Aiding public charities, agricultural and horticultural societies, public libraries, rifle associations, and benefit or superannuation funds for policemen, firemen and public employees, aiding railway companies, tramway companies, iron smelting works and the establishment of grain elevators.

15. Rewards for the discovery and apprehension of criminals, etc., etc.

16. The taking of a municipal census, and the keeping and returning of bills of municipality, etc., etc.

In addition to all these, local municipalities are empowered to establish, operate, or purchase waterworks, gas or electric light works, street railway or tramways and telephones, to acquire and maintain public parks and cemeteries, free public libraries etc.

The regulation and control of the sale by retail of intoxicating liquors, formerly entrusted to local municipal councils, has now been entirely withdrawn from them and vested in County Boards of License Commissioners appointed by the Provincial Government.

Municipal councils, however, still have power to increase the amount payable for shop and tavern licenses beyond the minimum fee fixed by the Government, and also to reduce the number of such licenses which may granted within the municipality.

In addition to the jurisdiction conferred upon municipal councils in regard to public health, very large powers relating thereto are exerciseable by the Boards of Health which exist in every local municipality.

In townships and villages, and in towns of less than 3000 inhabitants these Local Boards consist of the head of the municipality, the clerk thereof and three ratepayers appointed by the council for three years, one retiring annually; in larger towns and in cities, of the mayor and six ratepayers appointed by the council, two of whom retire each year. Two or more municipalities may by concurrent by-law unite to form an health district.

These local or district boards have power to regulate the ice supply of the municipality or district; to cause animals or articles offered for sale or intended to be used for food to be inspected, and if unfit therefor to be destroyed; to secure the isolation of cases of contagious or infectious disease, and if necessary to remove such cases to special isolation hospitals established by the Board or by the municipality, etc. The work of these Local Boards is further supervised and controlled by a Provincial Board of Health consisting of seven members appointed by the Lieutenant Governor-in-Council.

The duty of making and maintaining the public highways and bridges within a local municipality rests with the local municipal council; but county councils are responsible for the erection and repair of bridges

over rivers, streams, lakes and ponds forming or crossing the boundary lines of local municipalities, and of
bridges which cross streams and rivers more than one
hundred feet wide, and form part of a main county
highway.   County councils are also charged with the
maintenance and repair of any roads or bridges assumed
by the county with the consent of the local municipality in which they are situated. Default on the
part of any municipality in keeping in repair a road
or bridge within its jurisdiction, renders the corporation liable in damages to any person thereby injured,
but the action must be brought within three months.

For the purpose of keeping public highways in repair, every assessed owner or occupant of land in a
township municipality is required to perform annually
upon the highways a certain amount of "statute labour"
proportioned to the amount of his assessment, or in lieu
thereof to pay a commutation tax. In cities, town
and villages every inhabitant, and in townships every
non-resident land owner and every person not otherwise assessed pays an annual capitation tax of $1 in
lieu of statute labour.

The erection and maintenance of court houses and
land registry offices, gaols and houses of correction
belong exclusively to the councils of counties and of
cities; but with these exceptions, county and city
councils have but few powers which are not shared by
towns, especially towns which have separated themselves from the county municipality within which
they are locally situated.

In all cities, and in such of the larger towns as may
so determine, the appointment and control of the
police, and the regulation and control of children engaged in certain occupations is vested in a Board of
Commissioners of Police, which consists of the Mayor,
the County Judge and the Police Magistrate, the two
latter being appointees of the crown.

In cities of 100,000 inhabitants or more, (of which

Toronto is at present the only example) these commissioners further exercise, as to the licensing and regulation of various trades and occupations, all the powers which in other municipalities are exercised by municipal councils.

For the purpose of local taxation, the rateable property in each local municipality is annually assessed and valued by the assessors of the municipality, who, upon the completion of their assessment roll, are required to swear to its correctness and then to file it with the municipal clerk. Appeals against the valuations thus made are heard in the first instance by a Court of Revision composed (in townships, villages, towns and cities having less than 30,000 inhabitants), of five members of the local council. In cities of 30,000 inhabitants and over, the Court of Revision is composed of an official arbitrator or the sheriff, a member appointed by the city council, and a third by the mayor. The decisions of this Court of Revision are again subject to appeal, at the instance of any ratepayer interested, to the Judge of the County Court of the county in which the municipality is situated, or, (in the case of appeals involving large amounts) to a board of three County Judges.

The assessment roll thus finally revised becomes the basis of a "collector's roll" upon which the amount of taxes for general (local or county), special, and school purposes is entered by the municipal clerk; and the roll is then handed to the local collectors for the purpose of collecting these taxes.

If the taxes thus imposed cannot be collected from the property, or by distress upon the goods of the person who ought to pay the same, they become a charge upon the land. A schedule of all lands thus charged with taxes three years in arrear is annually returned by the local municipal clerk to the treasurer of the county or city, who thereupon offers them for sale by public auction.

QUEBEC.

The cities and most of the larger towns in this Province are incorporated by special statutes, which (in the case of towns) usually include by reference the provisions of the general code applicable thereto. The remainder of the province is subject to the provisions of a general Municipal Code, and is divided into:

(1) County municipalities (coterminous in most cases with the counties as divided for the purpose of representation in the Legislative Assembly).

Each of these may include several;

(2) Local municipalities, classified as (a) country or rural, (b) village, or (c) town municipalities.

A country or rural municipality may consist of a parish or part of a parish, a township or part of a township, or parts of more townships than one, but in order to be incorporated it must contain at least three hundred souls. A village municipality must contain at least forty inhabited houses within a space not exceeding sixty superficial arpents; and any village may be erected into a town by proclamation of the Lieutenant-Governor.

The affairs of each local municipality are managed by a council consisting of seven councillors elected annually (or appointed by the Lieutenant-Governor where no election has taken place) and holding office for three years. Of these, two councillors retire annually in each of two years and three in the third year. Local councillors receive no salary or remuneration for their services. Municipal electors must be males, twenty-one years old, British subjects, rated on the valuation roll of the municipality as owners of real estate in the municipality worth at least $50, or tenants, farmers, lessees or occupants of real estate worth $20 per annum, and must have paid by December 15th, preceding the annual election, all municipal and school taxes.

The elections for local councils are held annually on

the second Monday of January at the place where the local council usually holds its sessions, and are presided over in each municipality by a person appointed by resolution of the council, or, in default of such appointment, by the secretary-treasurer of the municipality *ex-officio*.

Any municipal elector who resides or has his place of business within the limits of the municipality, and who, in his own name or in his wife's name, is a proprietor of real estate worth at least $400, may be elected a local councillor. In the case of a rural municipality, any person domiciled in an adjacent village, town or city municipality, and otherwise qualified, may be chosen a member of the council of the rural municipality, provided he does not fill any municipal office in the municipality of his domicile.

Local councils at their first meeting choose from among themselves a mayor, who, in addition to the qualification necessary for the office of councillor, must be able to read and write. It is the duty of every municipal council to appoint a secretary-treasurer who is practically a permanent officer, and who, in the case of local councils, combines with his other duties those of tax collector. Local municipal councils are also required to appoint in every second year three valuators, a road inspector for every road division in the municipality, a rural inspector for each rural division, and as many public pound-keepers as they deem necessary. The valuators prepare every third year a "valuation roll" of all taxable property in the municipality, which, after examination by the local council, is forwarded to the council, and, having been equalized by them, becomes the basis of taxation both for local and county purposes for the next three years subject to annual revision by the local council.

The provisions respecting the sale of lands for arrears of taxes closely resemble those of the Ontario Act, except that the lands are sold in the year immediately succeeding default, and the time for redemption by the owner is two years instead of one year as in Ontario.

The powers of the local councils are much the same as those of the corresponding bodies in Ontario, but in addition they are further authorized in Quebec to prohibit as well as to regulate the sale by retail of intoxicating liquors within the municipality. As in Ontario, by-laws creating debts require, before coming into force, the approval of the electors whose property will be subject to the rates thereby imposed. Any person interested may appeal to the county council from the passing of any by-law by the council of a local municipality except (*a*) by-laws relating to the sale of intoxicating liquors and (*b*) by-laws which require the approval of the electors.

The county council is composed of the mayors of all the local municipalities within the county. It meets quarterly in March, June, September and December, and at the March meeting chooses from among its members a presiding officer styled a warden. In addition to the powers of a local council, the county council has also authority,

(1) With the concurrence of two-thirds of its members, to fix or change the *chef-lieu* of the county; but after a registry office has been established at the county town, the *chef-lieu* can only be changed by Act of the Legislature.

(2) To determine the place where the circuit court of the county shall be held, and to provide for the building of a court house and registry office, and for the purchase of the necessary land.

(3) To levy tolls on county bridges, and

(4) To prescribe the character of the vehicles which may be used in winter on municipal roads, or on turnpike roads belonging to trustees.

(5) The county council may further make provision for the payment of an indemnity to the warden and members for board and travelling expenses.

As in Ontario, local municipalities have jurisdiction over municipal roads within the municipality; county municipalities over roads forming municipal boundaries or partly in more municipalities than one, and (jointly) over roads forming county municipal boundaries; but county councils may by resolution or *procès-verbal* declare any local road a county road or *vice versa*.

In case a municipal council determines to open, widen, alter, or change the position of any public road or bridge within the municipility, they may either do so by means of funds raised by direct taxation, or by means of a *procès-verbal* drawn up by a superintendent specially appointed therefor, and which must specify (*a*) the work to be done and the time within which it is to be completed, (*b*) the properties liable to contribute, either by the labour of their owners or occupants or by supplying or paying for materials, (*c*) the proportion of the work to be performed by each ratepayer and (*d*), the person under whose superintendence the work is to be done. The *procès-verbal* being filed in the office of the council, is after examination and adoption (homologation) by them, (subject to an appeal by any interested ratepayer) followed by an Act of apportionment passed by the council, giving to it the force and effect of a by-law.

In respect of roads, bridges or other public works in which two or more counties are jointly interested, a Board of County Delegates consisting of three persons annually appointed by each county council (of whom the warden is *ex officio* one) may exercise the powers above recited in reference to such roads, bridges or other public works.

NEW BRUNSWICK.

In this Province there are no local municipal councils, the municipal affairs of each county being managed by a county council, composed of two councillors elected annually from each parish (a term which includes a

city or town as well as a rural municipality), except
the parishes of St. Stephen (which elects five county
councillors), Moncton (which elects three), and St.
John, where the council of the united city and county
consists of twenty-five councillors, of whom ten represent the city of St. John, and five the town of Portland. A county councillor must be an inhabitant of
the county, possessed of freehold or leasehold real property therein to the value of $500 above encumbran-
ces. The municipal elections are held annually in each
county on October 31st, or on such other day in October
as may be appointed by by-law—nominations for office
being made in writing. If more candidates than the
required number are nominated, a poll is at once opened,
and every male British subject twenty-one years old,
being a ratepayer of the parish on income or personal
property or both, to the amount of $100, or on real
property to any amount, and having paid his taxes on
or before the day of election, and whose name is on
the list of ratepayers who have so paid their taxes, is
entitled to vote.

The voting is by ballot, the ballots being counted
at the close of the poll by the chairman of the
meeting, assisted by tellers chosen at the hour of closing
by each candidate, or by some one on his behalf. The
county council meets semi-annually on the third Tuesday in January, and the first Tuesday in July, and at
the January meeting chooses a warden, who holds
office until the January meeting of the succeeding
council. Except in the county of Carleton, no warden
or councillor receives any salary or emolument.

The permanent officers of the county council are a
secretary and a treasurer, or a secretary-treasurer. The
county council at its January meeting further appoints
for each parish three overseers of the poor; two or
more constables; three road commissioners; collectors
of rates; a parish clerk; two or more fence-viewers;
two or more pound-keepers; a market clerk; two or

16

more hay-reeves ; boom-masters ; road surveyors ; dam surveyors ; surveyors of grindstones ; field drivers ; weighers of hay and straw ; measurers of salt ; measurers of wood or bark ; inspectors of barrels ; weighers of coal, and not more than three "by-road commissioners," who expend any moneys appropriated by the legislature for by-roads within the county ; also fire wards or wardens and wharfingers.

Property is rated (as in Ontario) at its actual value, and rates and taxes are levied as follows : One-sixth of the amount required is raised by a poll-tax on all adult males of twenty-one years of age and over, the remaining five-sixths being levied upon rateable real and personal property and income. The valuation for assessment purposes is made in each county by a board of county valuators, no two of whom may be residents in the same parish, city, or town. The valuators hold office for three years, and are apparently eligible for re-appointment. A valuation for taxation purposes is made every fifth year, based upon assessment schedules, furnished by the board of valuators to the assessors of each city, town, or parish.

The assessors, after having completed their schedules, make up a list of the names of persons assessed, and a valuation of the property and income of each, and post these lists in three of the most public places in the parish, requesting any person who objects thereto, to apply to the assessors for a reduction. After hearing such applications, the assessors may amend their schedules. If they decide against the applicant, he may appeal in writing to the valuators, by whom the assessment schedules are finally settled.

From the schedules so settled, the assessment roll, containing the names of the poll-tax payers and of the ratepayers on property and income, is made up and delivered to the tax collectors for the various . parishes, cities, or towns.

Taxes unpaid upon demand are collectable by execution in the nature of a *fi. fa.* goods, and if no goods can be found, then (in the case of persons residing within the province), by imprisonment of the defaulter for a term not exceeding fifty days ; in the case of non-residents within the province, by execution against the lands and the sale of so much thereof as a sheriff's jury of three freeholders not resident within the parish, may adjudge to be sufficient to pay the amount in arrear with the costs and expenses incurred.

NOVA SCOTIA.

Most of the cities and larger towns of Nova Scotia are incorporated by special statutes, *e.g.*, Halifax (in which the municipal council consists of a mayor annually elected by general vote, and eighteen aldermen elected for three years, six of whom retire annually), but several of the towns have been incorporated under a general Act (51 Vic. c. 1) now consolidated as the " Towns Incorporation Act." (58 Vic. c. 4.)

Under this statute any locality having 700 inhabitants dwelling within an area of 500 acres may, if the inhabitants of the locality so determine by vote, become incorporated by proclamation as a town, which is thenceforward governed by a town council consisting of a mayor and not less than six councillors.

A municipal elector must be a British subject, twenty-one years old, qualified to vote at elections to the Legislature, rated on the preceding year's assessment roll, and not in arrear for taxes of any kind.

A town councillor must be a British subject, and a resident ratepayer of the town for at least one year prior to the nomination ; a mayor must have been a resident ratepayer for at least three years and must be assessed on real estate for at least $500, or on personal property for at least $1,000.

The election of mayor and councillors is held on the first Tuesday of February of each year, the mayor being elected annually, and the councillors for two years, three of the six retiring annually. Nominations of candidates for the office of mayor and councillors are made in writing and handed to the town clerk at least a week before the day of election. The voting is by ballot upon voters' lists previously prepared by revisors appointed by the town council, the polls being held in each ward (if the town is divided into wards) and being open from 9 a.m. to 4 p.m. The procedure as to voting, counting of ballots, and return of the result to the town clerk resembles that prescribed by the Ontario Act, except that the official declaration of the election of the councillors for each ward is made by the presiding officer for the ward instead of by the town clerk.

The town council exercises all the jurisdiction formerly exercisable by the county council, the town meeting, the grand jury, the trustees of schools, the supervisors of public grounds, commissioners of streets and highways and overseers of the poor, and, in lieu of all county rates and assessments collects within the town and pays annually to the county treasurer its *pro rata* proporation of the expense of all services for their joint benefit; which proportion, in default of agreement, is determined by arbitration. The town council has also all the powers of a local board of health.

The council appoints a town clerk and all other necessary officers, including assessors.

Every male resident of a town between the ages of eighteen and sixty (except firemen, etc.), pays an annual poll tax of $2. In addition thereto, all real and personal property and income is liable to assessment and taxation as follows :—

The assessors make up their rolls annually and give notice thereof to every ratepayer by advertisement and also by personal service. Appeals from the assessment are heard by an " assessment appeal court," composed of three members of the council and the mayor, or the recorder of the town, who is an officer appointed by the town council, and must be a barrister. The assessment roll as thus revised becomes the basis of the " rate roll " according to which the town taxes are collected. Taxes unpaid, if uncollectable by distress become a lien on the land, which (when they are three years in arrear) may be sold by the town clerk therefor.

Outside the towns, the unit of municipal organization is the county or district corporation, the law as to which is contained in chapter 3 of the statutes of 1895. The council of a county or district is composed of one councillor for each polling district as laid off for the election of members of the Legislature. The election is held on the third Thursday of November in every third year, the nominations being made in writing ten days previous thereto and filed with the presiding officer of the polling district, who gives public notice thereof by poster or advertisement. The qualification of county councillors is the same as that required of members of the Legislative Assembly. Municipal electors are those qualified to vote for members of the Legislative Assembly, and the law governing provincial elections applies also to municipal elections. The voting is by ballot, the procedure being as above mentioned under the New Brunswick Act.

The council meet twice a year, the " annual " meeting being held on the second Tuesday in January (in Yarmouth on the third Tuesday), and the second " half yearly " meeting at such time and place as the council may by by-law appoint. Besides these regular meetings special meetings may be held at any time.

At the first January meeting after the election, the council choose from among themselves a warden, who holds office until the next election. The warden and members of the council may be paid then, travelling expenses with an extra fifty dollars to the warden. The county council appoints a clerk, treasurer, auditors, overseers of highways, road surveyors, constables, health officers, etc.

The matters in respect of which the county council may exercise jurisdiction are substantially the same as in the other provinces, but no by-law goes into effect until it has been approved by the Lieutenant-Governor in Council. No money can be borrowed without the authority of the Lieutenant-Governor in Council, and (if the amount required to be borrowed exceeds $2,000 in any one year) without the authority of an Act of the Legislature, to be applied for by the council after having first obtained the approval of the rate-payers at a town meeting duly convened for the purpose.

Chapter 5 of the Nova Scotia Statutes of 1895 consolidates the Acts relating to municipal assessments in counties. The district assessors make up their schedules annually between September 15th and November 1st, and forward them on November 16th to the municipal clerk. Appeals are made to a " Board of Revision," consisting of three rate-payers of the municipality, appointed by the county council at the January meeting. This Board meets on the 4th Tuesday in November and completes its duties by the 4th Tuesday of December of each year. On or before the 1st of April of the following year, each municipal clerk makes out from the amended roll the county rate and poor rate for his district, and forwards the roll to the district and township collectors.

The remaining provisions as to the collection of taxes, the sale of lands on default, etc., are substantially the same as in the Provinces of Ontario and Manitoba.

PRINCE EDWARD ISLAND.

In this province there are no municipal institutions in the usual sense of the term, i.e., no local municipal organizations, except in the two cities of Charlottetown and Summerside, both of which are incorporated by special statutes.

The Island being small, is practically (outside of these cities) but a single municipality, the affairs of which are managed by the Provincial Legislature.

" During every Session the House resolves itself into a Committee of the Whole, to consider all matters relating to public roads, and to pass resolutions appropriating money for this purpose in conformity with a certain scale arranged for the different townships. It passes Acts establishing and regulating markets, and making provision for the relief of the poor, for courthouses, gaols, ferries, roads and bridges, salaries, fire departments and various other matters which, in the larger provinces, are under the control of local (municipal) corporations " (Bourinot, " Local Government in Canada," p. 74).

There is indeed on the statute book an Act passed in 1870 (33 Vic. c. 20) which authorizes the Lieutenant-Governor to incorporate any town or village if two-thirds of the adult male householders resident therein so desire and express by a formal vote their wish to be so incorporated, but so far as I have been able to ascertain, this statute has hitherto remained a dead letter.

MANITOBA.

The county municipality, which in New Brunswick forms the unit of the system of local government, has no place in that of the Western Provinces. Municipal corporations in Manitoba are: (1) cities, (2) towns, (3) villages, (4) rural municipalities.

In order to become incorporated, a village must contain 500 inhabitants within one square mile, but if the population exceeds 2,000, 160 acres may be added for each additional 1,000 inhabitants. Towns contain 1,500 inhabitants for the first square mile, but if the population exceeds 2,000 then 160 acres may be added for each additional 1,000 inhabitants. Cities must contain at least 10,000 inhabitants.

The council of a city consists of a mayor annually elected, and of two aldermen for each ward, elected for two years, one of whom retires annually.

Town councils consist of a mayor and two councillors for each ward, annually elected, except in St. Boniface, where they are elected for two years, one of them retiring annually.

Village councils are annually elected, and consist of the mayor and four councillors; and the councils of rural municipalities (also annually elected) of the reeve and such number of councillors as may be determined by by-law, not being less that four nor more than six.

The qualification for municipal office is residence within the municipality and ownership of assessed real estate worth (over and above incumbrances);—in cities, freehold to $500 or leasehold to $1,000; in towns and villages, freehold to $500 or leasehold to $1,500; in rural municipalities there is no prescribed minimum of value.

The qualifications of municipal electors are the same as in Ontario, excepting that there are no income voters.

The nomination meetings in towns, villages and rural municipalities are held on the first (in cities on the second) Tuesday of December, and the elections on the third Tuesday in December, the proceedings thereat being precisely the same as in the Province of Ontario.

The Manitoba Act also provides for a recount of votes by a County Court Judge and for the trial of controverted elections by summary proceedings upon a petition presented to the Judge, from whom an appeal lies to the Court of Queen's Bench for the Province.

Bribery, undue influence, personation and fraudulent voting are punishable as in Ontario, by pecuniary penalties, and (in the case of a candidate) by disqualification for two years.

The officers of municipal councils and their respective duties are the same as in Ontario, except that the offices of clerk and treasurer may be held by the same person, who is then called secretary-treasurer, and who is required to be an adult male British subject.

The provisions of the law as to annual estimates, by-laws creating debts, assessments for local improvements, and municipal finance, are taken almost *verbatim* from the provisions of the Ontario Statutes. Special powers are, however, conferred respecting the drainage of rural municipalities, and as to providing a water supply therefor.

The legislative powers of municipal corporations are the same as in the Province of Ontario, and the Assessment Act of Manitoba is almost identical with that of the older Provinces.

THE NORTH-WEST TERRITORIES.

Municipal institutions in the North-West Territories are regulated by an ordinance of the Legislative Assembly passed in 1894 and amended in 1895 and 1896.

Municipalities may be either cities, towns or rural municipalities. Towns must contain over 500, and cities over 5,000 inhabitants.

The councils of all municipalities are elected annually, and hold office for a single year. The council of a city consists of mayor and ten aldermen; town councils, of the mayor and six councillors, but if the population exceeds 3,000, the number of councillors may be raised to eight; the council of each rural municipality consists of the reeve and four councillors. Any person is qualified for municipal office who is an adult male subject of Her Majesty, able to read and write, and (in cities and towns) resident within the municipality or within two miles thereof, and possessed of freehold or leasehold or partly freehold and partly leasehold real estate, assessed in cities for $2,000 if

freehold, and $5,000 if leasehold ; in towns for $500 if
freehold or $1,500 if leasehold ; in rural municipalities
of $400.

Municipal electors are men, unmarried women and
widows of full age assessed for

(A) Income.

(B) Personal property valued at $200, or

(C) As occupants or owners of real property to the
same value ; and all electors must have paid their
municipal taxes for the preceding year.

Nomination meetings are held on the first Monday in
January, and elections (*i.e.* polling) on the Monday of
the following week, the proceedings thereat being
precisely the same as in the Province of Ontario and
Manitoba, as are also the provisions of the law respecting corrupt practices and controverted municipal
elections.

The legislative powers of municipal councils are
taken from those of the Ontario Statute, with only
such amendments as the altered circumstances of the
Territories suggest.

The municipal officers are the clerk, the treasurer,
the auditors, the road overseers, their duties being the
same as in the other Provinces. The provisions of the
Statute respecting the assessment and collection of taxes
are based upon those of the Ontario Statute, except that
in rural municipalities an assessment need not be made
annually, but the council may decide that an assessment roll shall be used for three years consecutively
for the purpose of striking rates. Councils have
also power to levy taxation exclusively upon land
values without improvements. In other respects the
procedure relative to the assessment of lands, the collection of taxes and the sale of land for taxes two years
in default is taken from that prescribed by the Ontario
Municipal Act.

BRITISH COLUMBIA.

Although the number of municipalities in this province is small, the provisions of the law relative thereto are very complete, and closely resemble those of the older provinces.

There are two classes of municipalities,

1. Cities or towns, which must contain at least 100 adult, male British subjects, resident within an area not exceeding 2,000 acres, and

2. Township or district municipalities, containing at least thirty male adult British subjects who have resided within the municipality for at least six months.

The councils of all municipalities are elected annually. Those of cities consist of a mayor and not less than five nor more than nine aldermen; those of township or district municipalities, of a reeve and not less than four or more than seven councillors. The mayor of a city must be an adult male British subject, who for six months before nomination has been the registered owner of land in the city, assessed for $1,000 above encumbrances. Aldermen, (except in Nanaimo), must be similarly qualified upon an assessment of $500. In the city of Nanaimo the property qualification for the mayor is $500 and for an alderman, $350. In township or district municipalities the reeve must be a male British subject, registered as the owner of land in the municipality worth $500 above encumbrances. A councillor must be a male British subject, registered as

(A) The owner of land in the municipality assessed at $200 above encumbrances, or

(B) A homesteader or pre-emptor resident in the municipality for a year prior to the nomination and assessed for $500 above encumbrances.

Municipal electors are adult British subjects (males or females) resident within the municipality for a year prior to the nomination and not in default for municipal taxes, and who are:

(A) Owners or occupants of lands in the municipality, or

(B) Who carry on therein a business paying an annual trade license fee of not less than $5, or

(C) Householders within the municipality paying a rent of not less than $60 per annum. Chinese, Japanese and Indians are not entitled to vote at municipal elections.

The nomination meetings are held on the second Monday in January of each year, and the polling, if any, on the following Thursday.

Votes are given by ballot, and the procedure at municipal elections follows closely that prescribed by the Ontario Statute, as do also the provisions of the law respecting controverted municipal elections, and the punishment of bribery, treating, undue influence and fraudulent voting.

The sections of the statute relating to the legislative power of municipal corporations in British Columbia are in great part taken from those of the Ontario Act, but in addition to the clauses of that Statute, provision is made in this Province for raising revenue by levying trade licenses upon persons carrying on the business of selling liquor by wholesale or retail, restaurant keepers, keepers of billiard tables, bowling alleys, rifle galleries, opium dealers, etc. and also upon wholesale and retail merchants or traders, bankers, lawyers, land agents, express companies, gas companies, telephone companies, electric light companies, street-railway or tramway companies, investment and loan societies, etc. Municipal by-laws only come into effect after publication thereof in the Official Gazette, and in one or more newspapers circulating in the municipality.

The mayor or reeve is the chief executive officer of each corporation, and appoints the standing committees thereof. He has also power to veto or to return for reconsideration, any by-law, resolution or other proceeding of the council which has not already been reconsidered by them. Upon such return, the council must reconsider the by-law, etc., with the amendments, suggestions or objections of the mayor or reeve, and may thereupon either pass or reject the measure, which, if so rejected, cannot be re-introduced during the year, except by the unanimous consent of both mayor and council. The council have power to appoint by resolution or to elect by ballot such officers of the municipality as they consider necessary.

If the electors of the municipality so determine, the Lieutenant-Governor may appoint a special municipal auditor with very extensive powers of control over the expenditure or contemplated expenditure of the municipality, subject to an appeal to a Judge of the Supreme Court of the Province.

The assessment rolls are prepared and revised and the tax collection carried out in the same manner as in the Province of Ontario, lands in default for taxes being sold by the collector.

PART III

THE ECONOMICAL RESOURCES, TRADE AND POPULATION OF CANADA.

PART III.

CHAPTER I

The Fur Trade of Canada.

By Sir Donald A. Smith, G.C.M.G., High Commissioner of
Canada, and Chairman of the Hudson's Bay Company.

THE fur trade and fur-traders have always been prominent features in Canadian history. Even in the
days of the French régime the business was of importance, and the leading occupation of the people. It led
to the exploration of the country, and to the discovery
of the great waterways which have had so much
influence on the development of Canada. The fur-
traders were undoubtedly the pioneers of civilization
in North America. Long before the English penetrated inland, the French *coureurs de bois* had preceded them, and had established relations, more or less
friendly, with the Indians. It was the fur trade which
led to the formation of the Hudson's Bay Company in
1670, or to give its proper title, "The Company of
Adventurers of England trading into Hudson's Bay."
In 1666, two French explorers journeyed to what is
now known as Manitoba, and made their way by
Lake Winnipeg and the Nelson River to Hudson Bay,
which had been discovered by Hudson on his memor-

able voyage fifty-six years previously. On their return to Quebec, they offered to conduct ships by water from that city through Hudson Strait to the heart of the fur country, avoiding the long and tedious canoe journey by way of Lake Superior. Their proposition was not, however, entertained by their countrymen. At the instance of the British Ambassador in Paris at that time, they conveyed the knowledge they possessed to London. An expedition of the nature they suggested was organized, and, being successful, led, as already mentioned to the incorporation of the Company which has continued in existence ever since. For many years the trade was carried on at five or six forts or trading posts on the shores of the Bay. The collection of the furs has always been largely in the hands of the Indians, and the peltries were taken by them to the posts, and exchanged for the merchandise, supplies and trinkets, that were offered in return. During the struggle between France and Great Britain for the possession of North America, the Hudson's Bay Company suffered many losses and inconveniences, and the forts on Hudson Bay changed hands on several occasions. While the Englishmen were confining their operations to Hudson Bay, their French rivals took the bolder course, and established stations in different parts of the interior. They were thus able to intercept the furs, and in the end the Company was forced to follow the example of its competitors and go for the furs instead of waiting for them to be brought to the posts.

Even after France ceded her rights in North America to Great Britain, competition to which the Company was subjected still continued, but it then proceeded nominally under British or Canadian auspices. In the first place the opposition was conducted by individuals, but afterwards by associations. There was for instance the North-West Company, founded in 1773-74; and its off-shoot the "X. Y. Company," followed it in 1798.

They subsequently, however, joined forces and made a
most powerful competitor to the Hudson's Bay Company. The rivalry between the two bodies continued
for many years, until their amalgamation in 1821, and
resulted in much friction and bloodshed. In fact, on
more than one occasion, the attention of the Imperial
and Canadian Governments was called to the matter.
In the meantime the country had been explored north,
south, east and west. Stations had been established in
Labrador ; on the shores of the Gulf of St. Lawrence ;
along the Ottawa River and in the district north to
Hudson Bay ; west and north-west of Hudson Bay,
even to the confines of the Arctic Ocean ; and over the
Rocky Mountains into British Columbia. At the time
of the amalgamation there were probably about 140
posts in operation. It is not too much to say that it is
owing to the fur-traders that much of Western Canada
was retained for the British. This is particularly the
case in regard to what is now British Columbia, and it
is not their fault that the country between the Columbia River and the international boundary is at the
present time under another flag.

In the early days, Moose Factory, the headquarters
of the southern department, and York Factory, the
headquarters and port of importation for the northern
department, both on Hudson Bay, were the chief posts
of the Company. There, the furs were collected and
shipped, and supplies were sent from those places each
year to the most distant stations. In many of the districts the life of the officers of the Company was one of
isolation and solitude. The same thing applies with
almost equal force to the more remote posts at the
present time, although there are more travellers than
formerly. Still the life must have its attractions, for
the officials have always been devoted to their posts
and to their work. It was only on rare occasions that
they paid visits to civilization. They were alone for
years together, without any society but that afforded

by the employés themselves, and by those at the nearest station, often hundreds of miles away. Their object in life was to cultivate the most friendly relations with the natives, and to collect as many furs as possible both from the Indian and half-breed hunters and trappers. The great event of the year was the arrival of the canoes or dog-trains with supplies, and mails and packets of newspapers, magazines and books, which kept them informed of what had taken place in the outside world many months previously. These means of communication with the head office of the Company very rarely occurred more than once in the year, and the conveyances which brought in supplies, took back the furs that had been gathered. The Company's powers of exclusive trading in the Indian territory, which included the country west of the Rocky Mountains (obtained after the amalgamation with the North-West Company in 1821), expired in 1859, as did also their lease of Vancouver Island; and they surrendered the remainder of their rights and privileges to the Imperial Government ten years later, for (1) a payment of £300,000; (2) the reservation of certain lands about their forts and trading posts, and (3) one-twentieth of the land in the fertile belt as it might be surveyed. The fertile belt includes the country between the North Saskatchewan River and the international boundary. It might have been expected that the fur trade of the Company would diminish with the advent of civilization, and the construction of the Canadian Pacific Railway. This has turned out to be the case so far as the districts along the line of railway and in the zone of settlement are concerned, but the bulk of the trade has not largely fallen off, although the returns of the various posts have somewhat changed. The opening up of the country has made the trade more active in some of the less accessible regions, and the railway now plays an important part in the shipment of the furs, and in the

distribution of supplies Dog trains and canoes are
not so much used as formerly. except, perhaps, in the
more remote portions of the Dominion. Where the
railways are not available, the rivers and lakes are
utilized, and steamers are employed wherever it can
profitably be done. In addition to the trade with the
Indians in the interior, the Company now transacts a
considerable business in the leading towns and villages
in Manitoba, the North-West Territories, and British
Columbia ; and its stores, at which all the requirements
of the country can be met, will compare favourably
with similar establishments in the United Kingdom.
It is also a large landed proprietor ; at the present time
its land grant exceeds 3,000,000 acres, which may be
more than doubled as surverys progress. This land
will become very valuable as settlement proceeds. At
the present time Winnipeg is the chief collecting and
distributing centre for the North-West and western
trade. Montreal acts in a similar capacity for the districts along the Ottawa River, the vicinity of Lake
Huron, Lake Superior. etc., the posts on the Gulf of
St. Lawrence, and partly for the Labrador posts.
Moose Factory is still the port of entry and the headquarters of the southern department, but York Factory
which was formerly the depôt for the supply of the
whole of the southern department, is now only the
headquarters of the fur districts around the northern
and western portion of the Bay. A sailing vessel proceeds annually to Moose Factory from London, and a
steamer to York Factory, which calls on the way out
at the Labrador posts, but the bulk of the export and
import trade of the Company is done *via* Montreal.
The headquarters of the Company in British Columbia
are at Victoria. the supplies until quite recently having
been sent to that place, and the furs collected in the
interior sent there and forwarded to Montreal, by rail,
but Vancouver is now also becoming an important
depôt, as being in some respects more convenient for

the distribution of supplies than Victoria. The total shipment of furs varies with different years, as do the prices obtained for them. On the average, however, they realize at the auction sales in London somewhere between £200,000 and £300,000 per annum. In the early days of the Company the only currency and unit of exchange was the beaver skin. The price of everything was fixed on that basis. At the trading posts the Indians received in return for their furs pieces of sticks, specially prepared, each of which represented in value a beaver skin. They exchanged these at the stores for their value in supplies. Later on the Company started a paper currency, which became known as the Hudson Bay Blankets. The notes issued were chiefly of three denominations, £1, 5 shilling, and 1 shilling. Now of course in the modern establishments the currency of Canada prevails. The following are the principal skins which find their way to the London market from the Company :—Badger, bear, beaver, deer, fisher, fox, lynx, marten, mink, musquash, otter, rabbit, racoon, skunk, wolf and wolverine, the bulk, so far as numbers are concerned, being made up of beaver, marten, mink, and musquash.

Naturally a Company with the exclusive rights and privileges of the Hudson's Bay Company made enemies. It cannot be questioned, however, that it performed valuable work in exploring the country, and in preparing the Indians for the advent of the white man, and of settlement. Its administration made the inclusion of Western Canada in the Dominion a comparatively easy process. Its officers were always on good terms with the Indians, and it is to the wise policy adopted, and which was followed by successive Governments of Canada, that, so far as the Indians are concerned, the progress of the Western territory has been so peaceful. Except in a few isolated instances the Indians did not take part in the half-breed difficulties of 1869–70, or in those of 1885. During the

Hudson's Bay Company's régime. Western Canada was isolated and difficult of access. Not until nine years after its transfer to the Dominion did it receive railway communication, and it was sixteen years before it was placed in direct communication east and west with the Atlantic and Pacific coasts. Prior to Confederation, none of the Provinces of Canada could have taken over and administered the territory, nor could the Canadian Pacific Railway have been constructed. The verdict of history will undoubtedly be that the Hudson Bay Company performed useful functions, and that it played a most important part in the events which led up to the Confederation of Canada, especially so far as regards the western parts of the Dominion.

CHAPTER II.

The Fisheries of Canada.

By Professor Edward E. Prince, B.A., F.L.S., Dominion Commissioner of Fisheries, Ottawa.

O F the world's great fisheries those of Canada, are without doubt, the most vast in extent, and the most varied in their products. The waters on the Pacific and Atlantic shores of the Dominion teem with fish of the greatest economic value, while the system of fresh-water lakes, really inland seas, the lakelets, countless in number, and the noble rivers which flow through her far-reaching territory, provide the amplest field for gigantic fishing industries.

The annual value of the inland and sea fisheries is estimated at not less than $30,000,000 (£6,000,000) and their growth has been phenomenal. In 1850 their value did not exceed $150,000 (£30,000). In 1852 the value was doubled, and reached $300,000 (£60,000). In 1859 the value rose to $1,407,000; ten years later (1869) to $4,376,526; 1876, $11,147,000; 1886, $18,677,288, and, as already stated, it must now reach about $30,000,000, including the value of fish consumed by the Indians, the Eskimo, and the settlers in remote districts of the Dominion. An army of fishermen, over 60,000 in number, possessing boats, nets and gear, valued at about $10,000,000, engage in these fisheries.

The following summary, suggested by the system of territorial regions which Sir William Dawson laid down in his "Ice Age," recognizes seven great divisions, each characterized by fisheries more or less distinctive.

1. The *Atlantic division*, from the Bay of Fundy to the coast of Labrador, embracing deep-sea and in-shore fisheries, cod, mackerel, haddock, halibut, herring, hake, lobster, oyster, seal and white whale (Beluga) fisheries. Annual value : $10,000,000 (£2,000,000).

2. The *Estuarine* and *inland waters* of the *Maritime Provinces* (Nova Scotia, New Brunswick, Prince Edward Island and Quebec), including fisheries for salmon, shad, gaspereaux (alewife), striped bass, smelt, and in the lakes, winninish or land-locked salmon, lake trout or lunge, maskinonge, etc., of the annual value of $2,500,000 (£500,000).

3. The *Great Lakes* and *tributary waters* :—Lake whitefish, great lake trout, lesser whitefish (called lake herring), sturgeon, pike-perch (dore or pickerel), black bass, brook-trout, maskinonge, pike and numerous carps, suckers, and catfish. Value : $2,000,000, (£400,000).

4. *Great North-West Lakes* (including Manitoba), yielding lake whitefish, sturgeon, pike-perch, tullibee (a peculiar lesser whitefish), pike and gold-eye (fresh-water herring). Value, including newly developed " caviare" and " sturgeon sounds" industries, $1,000,-000 (£200,000).

5. *Pacific Interior* or *Rocky Mountain Plateau*, comprising little-developed fisheries, land-locked Pacific salmon, lake whitefish, lake trout, river trout and numerous Cyprinoids, none of which are identical with eastern species. Annual value small and unrecorded.

6. *Pacific coast fisheries* which are almost un-worked, if the estuarine salmon fisheries be excepted. At least seven different species of Pacific salmon occur belonging to the genus Oncorhynchus, excluding *Salmo gairdneri*, the steelhead. Halibut, skill (black cod), oolachan (candle fish), anchovy, herring, smelt, and a great variety of other marketable fishes abound, but are not to any adequate extent utilized. Shark,

dogfish and whale fisheries exist, and there are limited oyster fisheries. Exclusive of the fur seal, which is an oceanic industry, less than $1,000,000 in value, the coast fisheries may be valued at $4,000,000 (£800,000).

7. *Hudson Bay* and *Peri-Arctic Area* (Ungava Bay to the Mackenzie River):—Whale, walrus, sea-trout, the inconnu, which is a huge river whitefish, pike, suckers, sturgeon, and possibly salmon and cod, occur in these vast waters, of which Hudson Bay alone exceeds the Mediterranean Sea in extent. The richest whaling grounds in the world are in this little-known part of Canada, off the mouth of Mackenzie River and as far east as Cape Chudleigh in Hudson Strait, where the Baleen whale and walrus were until recently numerous. "The tidal channels of Canada's Arctic archipelago are destined," it has been truly said, "to be the last home of the leviathans, which within the memory of living men have been driven from New-foundland latitudes to the places where their survivors have now sought retreat."

It may be pointed out that the waters grouped in this seven-fold manner include, on the Atlantic, a Canadian coast line at least 10,000 miles long, and on the Pacific not less than 8,000 miles, while the portions of the great lakes (Superior, Huron, Erie and Ontario) which lie within the British boundary line, embrace a fishing area exceeding 70,000 square miles. To these extensive waters must be added giant streams like the St. Lawrence, the largest river on the North American continent, the Mackenzie River (2,400 miles long), the Saskatchewan (2,000 miles long), the Fraser and Red Rivers, each 600 miles long, and others, like the Rivers Peace, Assiniboïne, Skeena, Ottawa, St. John, Restigouche and Miramichi, all of which are great rivers, abounding in the choicest species of fish.

It is important to note that the Atlantic inshore fisheries of Canada are prosecuted not by Canadian fishermen alone, but by those of the United States,

Newfoundland and France under International Treaties. The great lakes also are for the most part divided between the United States and Canada, and the recorded Canadian catches represent therefore only a proportion of the total yield of those waters.

In Hudson Bay and the northern seas as well as in the Pacific inshore waters of British Columbia, foreign fishermen have very largely encroached on the fishery resources of the Dominion. There are, it may be added, extensive waters as yet untried and undeveloped, and valuable resources which in the near future will add to the annual value of the Canadian fisheries.

The importance of fishing industries did not in the past go unrecognized. A Government Department charged with the administration of Fishery, as well as Shipping matters, was created at Confederation (1867), prior to which, the fisheries had been regulated for nearly thirty years by a branch, organized in 1859, of the Crown Lands Department of Upper Canada. Such control as the Provincial Governments still exercise in Ontario, Quebec, and the other Provinces, is carried out by the Commissioners of Crown Lands in the several Provinces. Since Confederation the vast fisheries of the Dominion have been under the direct supervision of a Cabinet Minister (the Minister of Marine and Fisheries) at Ottawa. A Deputy Minister acts immediately under the Minister, and a Commissioner of Fisheries, who is also General Inspector for the Dominion, has important advisory and executive functions. In addition to the usual inside staff of clerks, a body of outside officers enforce at a yearly cost of about $120,000, the close seasons, and the fishery license system, as well as collect statistics, etc. The staff includes twelve Inspectors of Fisheries (who receive $700 to $1,500 per annum); several hundred Overseers vested with magisterial powers for the purposes of the Fisheries Act (receiving $100 to $900); and a still larger body of temporary fishery guardians whose pay ranges from $1.50 to $2 per day.   A fleet of armed cruisers, costing about $100,000 annually, patrol the coastal and great inland waters, exercising surveillance over foreign as well as Canadian fishing operations in Dominion waters.   Finally, a Bounty System is carried out for encouraging the pursuit of the deep-sea fisheries in the Atlantic, the provision for which was secured by the Halifax Award (November 23rd, 1877), whereby a sum of $5,500,000 was paid by the United States in consideration of the fishery concessions granted to the United States fishermen in Canadian inshore waters along the Atlantic coasts. A sum of $160,000, voted annually by Parliament, is by this means available, and is distributed amongst the deep-sea fishermen in the Maritime Provinces. The work of the Fisheries Department is thus extremely varied and important.   The late Professor Brown Goode, United States Commissioner of Fisheries, at a Fisheries Conference in London, 1883, said :—" It seemed to him that the Canadian Department of Marine and Fisheries was one of the most valuable organizations in the world, and that the system of gathering statistics was one which other countries ought to study with a great deal of care.   In the United States they had nothing of the kind."   The collection and publication of statistics is, indeed, an invaluable branch of the Department's work.

The methods of protection and restoration adopted by the Department of Marine and Fisheries are :—

1. Close seasons preventing the capture of spawning fish.

2. Fishing licenses specifying the amount of net, kind mesh, etc.

3. Prohibition of obstructions, pollutions, etc.

4. Artificial fish culture as a means of introducing fish into new waters, and supplementing natural reproduction.

The last is carried on by means of fourteen hatcheries

under the supervision of the Commissioner of Fisheries. Salmon (Atlantic and Pacific), great lake trout, and lake whitefish, are hatched and shipped *gratis*, if the waters applied for are suitable. A lobster hatchery at Picton, N.S., turns out annually one hundred millions to one hundred and sixty millions of minute larval lobsters. The fish culture operations cost between $30,000 and $40,000 per annum, and in 1895, close upon three hundred millions of the fry of the various fishes above named were planted in the several provinces.

A sea-fisheries Intelligence Bureau, established in 1889, including between fifty and sixty stations under the charge of the Commander of the Protection Fleet announces daily to the fishermen the movements of fish and the localities for bait.

The following table shows in graduated series the various fish and fish products with the relative value of each :—

Kinds of Fish.		1894.	
		Quantity.	Value.
			$ cts.
Cod .... Cwt.		938,027	4,225,896 00
" tongues and sounds .... Brls.		833½	8,335 00
Salmon, preserved in cans ...... Lbs.		23,647,162	2,365,717 30
" fresh .... "		5,484,653	801,429 80
" pickled .... Brls.		5,629	51,404 00
" smoked .... Lbs.		80,280	8,888 00
Lobsters, preserved, in cans...... "		13,333,693	1,804,256 66
" in shell, alive, etc..... Tons.		7,565	567,375 00
Herring, pickled................ Brls.		439,238	1,977,336 00
" fresh or frozen .... Lbs.		16,966,241	404,965 86
" smoked .... "		9,100,980	183,427 60
Whitefish .... "		14,854,170	879,650 46
Mackerel, pickled .... Brls.		53,087	731,782 00
" fresh and preserved.... Lbs.		1,803,072	177,088 14
Trout .... "		7,182,083	720,906 80
" pickled .... Brls.		3,724	37,240 00

Kinds of Fish.		1894.	
		Quantity.	Value.
			$ cts.
Haddock, dried	Cwt.	137,140	479,987 50
" fresh preserved, etc....	Lbs.	503,490	36,559 20
Smelts	"	8,087,079	404,882 95
Hake	Cwt.	103,297	263,059 00
" sounds	Lbs.	83,187	41,593 00
Pollock	Cwt.	88,758	221,894 00
Halibut	Lbs.	3,481,276	254,151 90
Alewives	Brls.	63,470	253,904 00
Pike	Lbs.	3,079,484	81,655 75
Sardines	Brls.	136,828	274,756 00
" preserved	Cans.	220,000	11,000 00
Pickerel	Lbs.	7,610,425	293,266 25
Oysters	Brls.	45,127	182,108 00
Sturgeon	Lbs.	2,182,071	119,055 10
Coarse and mixed fish	Brls.	73,167	226,373 95
Eels, pickled	"	7,978	75,116 60
" fresh	Lbs.	951,350	48,979 32
Bass	"	1,289,461	93,800 86
Shad	Brls.	9,244	92,432 30
Tom-cod or frost-fish	Lbs.	1,816,320	90,815 50
Clams			62,996 25
Squid	Brls.	14,868	59,470 00
Maskinonge	Lbs.	627,457	37,647 42
Mixed fish (British Columbia)			24,693 00
Flounders	Lbs.	424,320	20,975 50
Crabs			18,000 00
Oulachons	Lbs.	336,700	17,090 00
Winninish	"	100,000	6,000 00
Fur seal skins in British Columbia	No.	94,474	944,740 00
Hair seal skins	"	21,643	25,405 00
Sea otter skins	"	12	1,500 00
Porpoise skins	"	97	388 00
Perch	Lbs.	971,814	28,970 08
Fish oil	Galls.	745,848	298,338 40
Fish used as bait	Brls.	250,984	332,417 00
" manure	"	106,239	53,120 00
Fish guano	Tons.	5,117	71,525 00

CANADIAN FISHES.

It is necessary to add a few succinct notes upon certain species of fish of prime importance, commercially,

or for sport, which are either peculiar to the waters of
this continent or closely allied to European species.
The cod, haddock, halibut, mackerel, herring, salmon,
pike-perch or doré (also called pickerel), the pike,
smelt, eel, and other kinds, call for no special reference ;
but others, like the whitefish, striped bass, etc., demand
a brief notice.

Whitefish (*Coregonus clupeiformis*, Mitchill). This fresh water salmonoid is allied
to the European gwyniad and pollan. It
varies in weight from 2 lbs. to 16 lbs., and is deep in
the body, the shoulder abruptly descending to the
head, which is very small, the jaws are toothless, the
snout blunt, and the gape contracted. The large silvery
scales upon its sides, or as some think, the whiteness of
the flesh have gained for it its distinctive name. No
fish is more justly esteemed for table purposes, and
to explorers and Indians it is invaluable as a continuous
diet of whitefish unlike salmon never palls upon the
taste. There are several species abounding in almost
all the lakes from the Atlantic to the Pacific, and their
capture constitutes one of the most valuable of the
fresh-water fisheries, the annual yield being not less
than nine or ten thousand tons, or about one-fifth of
the yearly take of codfish.

Lake Whitefish.

The lesser whitefish, called cisco and lake
herring, have become valuable in recent
years, as the larger species have been considerably depleted. They feed upon insects and small
crustaceaus, and like the Salmonidæ generally, they
resort in the fall to their accustomed spawning
grounds, traversing great distances, in many cases, to
do so.

Lesser Whitefish.

The speckled trout or brook trout of
Canada (*Salvelinus fontinalis*, Mitchill),
is more allied to the charrs than to the
common river trout (*Salmo fario*, L.) of Europe. Instead of the silvery sides with comparatively large

Speckled Trout.

scales, showing minute red and black spots, the
Canadian speckled trout has small scales, dusky green
back and dorsal fin vividly diversified with yellow
vermiform markings, the sides being spotted with
red, white and black. The reddish paired fine show a
cream-white anterior margin. It is more important
for sport than commercially, but its game qualities
are inferior to those of the English trout.

Maskinonge. The maskinonge (*Esox nobilior*, LeSuer),
bears a general resemblance to the pike
(*Esox lucius*, L.), but is in many respects superior.
Its edible and game qualities are remarkable, and it
often attains a weight of 70 lbs. Whereas the pike is
blotched with white on its greenish-brown or dusky
sides, the maskinonge exhibits brown blotches on a
pale ground colour. The branchiostegal rays are 17 to
19 in number, but in the pike 14 to 16.

Most of the still waters of Quebec and Ontario
contain this fine game fish, but it has greatly decreased
in numbers, though splendid fishing is still to be had in
Lakes Scugog, Rice, Simcoe, and other Ontario waters.

Black Bass. Black Bass (*Micropterus*, Lacep). The
two species of black bass rank high in the
estimation of the angler. They range from 2 lbs. to 8
lbs. and are bold, strong and of game qualities. The
flesh is firm, white, and of great excellence. The nest-
building habits and strong parental instincts of these
fish are well-known.

Striped Striped Bass (*Roccus lineatus*, Bloch.),
Sea Bass. occur in the tidal waters along the Atlantic
coast. They reach a great size (15 lbs. to
40 lbs.), and afford splendid sport. They are, with the
exception of the salmon, the choicest of food fishes,
but their destruction when dormant in the rivers in
winter, and the taking of the immature young in smelt
nets, has seriously depleted them.

Catfish. Catfishes or Siluroids (*Ameiurus*). A
great variety of species occur in the rivers
and lakes, and all are characterized by the long feelers

which project from the upper and lower jaws. In
size they range from 2 or 3 inches to 4 or 5 feet, and
as there is a good demand for them in the United
States markets considerable catfish fisheries have
grown up in some localities.

Gar-Pike or   Lastly, two remarkable species of fish
Bill-Fish,   call for notice on account of their scientific
Sturgeon,   interest, for commercially they are value-
Shad, etc.   less.

The two Ganoids, *Lepidosteus* (gar-pike) and *Amia*
(Bow-fin or Lake Dog-fish), are fairly plentiful in the
lakes and slow streams, especially in Ontario.

Of the sturgeon, shad, and the remarkable salmonoids
of the Pacific waters, it is not necessary to add any
remarks in this necessarily fragmentary sketch.

METHODS OF FISHING.

At least a dozen methods of taking fish for the markets, on account of their importance, merit a passing
notice, but the two chief methods are the pound-nets
or fish-traps and the gill-nets or drift-nets. The latter
(gill-nets) hang like a wall in the water, suspended by
floats, and the fish, in their endeavours to pass through,
become meshed by the head and strangled. The
former (pound-nets or weirs) consist of a "leader"
which obstructs the fish and leads them into a staked
enclosure, out of which they do not escape. Pounds
of wickerwork or brush are used in New Brunswick
and Nova Scotia for taking sardines, herring and
mackerel. Swing-nets and other forms of stake-nets
are used for salmon, etc., but the hoop-nets (or verv-
eaux) are perhaps the most widely used, for taking
the inferior kinds of fish, catfish, suckers (cyprinoids),
perch and the like. The hoop-net has the form of a
funnel held open by a series of erect wooden hoops and
usually set in creeks and inshore waters. A special
form of trap or weir is used for taking eels.

The seine is a most effective net, but on account of

18

its destructive nature, its use has been discouraged.
To the extensive use of seines in former years may be
largely attributed the serious decline in some localities
of once prolific fisheries. Scoop-nets and bag-nets are
used for taking smelts, striped bass, and shad. They are
effective when used through the ice, in winter, taking immense quantities of fish, carried in with the tide, as in
the case of the smelt, or lying torpid, like the striped
bass in the winter months.

Salmon and
Lobster
Canning.

The vast salmon and lobster canning
industries of Canada (salmon on the Pacific
coast, and lobsters on the Atlantic coast),
are in some respects the most remarkable fishery
enterprises in the world. Probably nine to ten millions
of salmon are annually used in British Columbia,
while every year from eighty to one hundred millions
of lobsters are packed in the six or seven hundred
lobster factories on the coast of New Brunswick,
Prince Edward Island, Quebec and Nova Scotia.

Oyster
Fisheries.

Finally, the oyster, which differs from the
European species in being diœcious and
in its hundred-fold more prolific character, is distributed over vast areas along the Atlantic coast, constituting these areas most extensive and valuable
oyster grounds. The annual take 50,000 to 70,000
barrels, represents but a tithe of the possible yield,
were systematic culture and judicious fishing methods
adopted.

CONCLUSION.

Ever since the discovery of this vast western continent the richness and value of the Canadian fisheries
have been acknowledged, and though the fishing fleets
of Norway, Portugal, Spain, France, and England,
have for centuries prosecuted commercial fishing in the
waters of the Dominion, and the old Colonial Provinces,
the United States and the British Provinces have
taken from them incalculable quantities of fish food
for the markets of both hemispheres, they still remain
the greatest and most varied fisheries in existence.

CHAPTER III.

The Forests of Canada with their Distribution.

By Professor John Macoun, Botanist of the Geological Survey of Canada.

THE forests of the Dominion of Canada are one of its chief assets and one that it seems the aim of governments and individuals to annihilate as quickly as possible. Instead of attempts being made to conserve these natural coverings of the land, means, both legitimate and illegal, have been taken to destroy them. In all the older provinces this has been done to such an extent that in many sections that were covered with unbroken forest fifty years ago there is to-day scarcely a tree to be seen. The great fertility of the land in former times is spoken of as if it pertained to the forest alone when in reality it was due to other causes.

On the sea coast, cutting away the forests has let in the sea air, and to-day the soil of Prince Edward Island and parts of Nova Scotia is wetter than when the timber was first cut off. As a proof of this, tamarack is now growing in pastures and meadows where hardwood once covered the land, and under-drainage has become an absolute necessity.

On the other hand the deforesting of Ontario has dried up springs, lessened the flow of rivers, caused sudden and early thaws in winter, and in summer droughts over large areas, and as a result lessened the products of the soil at least one-half. Year by year this state of things is becoming more intensified, yet the supineness of the authorities is so great that no sensible attempt is made to remedy this state of things. The forests of northern Ontario are being cut down to supply the increasing demand for pine and other woods, and in the wake of the cutting follows the annual fires which, besides burning over the districts from which the timber has been cut, extend in many instances through the untouched forests and destroy more timber than the woodman with his axe. Year after year this goes on, and now when a hundred miles or more intervene between the settlements and the lumber camps, little attention is paid to the subject, but when the public awakes to the truth it will be appalled at the enormous waste and loss that has been going on for more than a generation.

Twenty-five years ago the Algoma district, over 1,000 miles from east to west and we may say 200 miles from north to south, was a solid coniferous forest. To-day most of it is so completely denuded of trees that even the dead and whitened trunks of some localities have disappeared and nothing is to be seen for miles but bushes and young trees growing in the crevices of the naked rocks, repeated fires having burned up every particle of the former covering which was the accumulation of ages.

Any traveller going west on the Canadian Pacific railway from Ottawa will pass through 1,200 miles of what was once continuous forest. At present, he will see little else but a dreary wilderness of bare rock, burned and bleaching trunks or young forests trying to cover up the nakedness of the land. I am not citing the line of the Canadian Pacific railway as the particular line but only as an illustration, for there is no disguising the fact that any line will do. In the summer of 1868 the first opening was made in the forest at Port Arthur. The summer of 1870 saw Wolseley's expedition pass on its way to Winnipeg and that summer the forest at Port Arthur was burned and since then the havoc has been continuous.

There was a time when the prairie region was being deforested at an enormous rate and every year fires rushing from the south and west forced their way into the still untouched woodlands and extended the burnt area still farther to the north.   As soon as settlement took place attempts were made to stop the fires, and of late years destruction from that cause has almost ceased.   It is a fact, nevertheless, that at the time of Palliser and Hynde's expeditions in 1857-59 there were districts south of Qu'Appelle covered with heavy forests of aspen that twenty years after, in 1880, I found without even a twig to show that a tree ever grew there.

Passing westward to the Rocky and Selkirk Mountains, the same tale may be told.   Forests of tall, graceful trees invaluable for railway and other purposes filling the valleys and climbing the mountain sides in 1885, nearly all gone in 1893.   When the right of way was cut through the mountains, a lane was made through the forest and the brush and logs piled on either hand.   The burning of this started the fires that prepared the material for succeeding years when the fires climbed the mountains so that at Hector and Stephen on the summit of the Rocky Mountains not a green tree was to be seen in 1890 where they had stood in myriads in 1885.   This was not all, in 1885 quantities of permanent ice and snow that had completely disappeared in 1890, lay on the mountains to the north and south and instead of the cool mountain slopes of six years before the ascent had to be made through a blackened forest where the rustling of the dead bark and the tapping of the woodpecker took the place of the songs and twitterings of the small birds seen in 1885.

The same year the Columbia Valley from Golden down to Donald, and up Beaver Creek and down the

Illicilliweat to Revelstoke was an unbroken forest of tall stately trees ; to-day those that are left are ragged, torn and shrivelled, and the forest beauty has departed for ever. Year after year the lumberman is penetrating the valleys and the fire following in his wake finishes what he begins. In a few short years desolation will reign, and the avalanche that descended in the form of snow will be replaced by rivers of mud, trees and rocks. The mountains will be disfigured, and travelling in spring will be both uncertain and dangerous.

Each succeeding summer on Vancouver Island the same destruction goes on. A great deal of the interior has been burned over repeatedly, and owing to the long summer droughts and the lack of brush amongst the tall trees the moss and logs become dry and the fire when once started never ceases until the September rains commence when the air clears of smoke, the fires die out and all things remain soaked until the following July when the same round of fires begins again.

SUB-ARCTIC FOREST BELT.

Lying south of the watershed in Labrador and south of a line drawn north-westerly from Fort Churchill to near the mouth of the Mackenzie River in the North-West Territories is a belt of forest that is continuous except where the surface becomes a peat bog too wet to support trees or the depressions are deeper and become lakes. This extensive belt at the base of the Rocky Mountains extends from lat. 53 to 67 in the valley of the Mackenzie. It trends to the south as it goes easterly so that in the meridian of Lake Winnipeg its limits are between 50 and 58 ; passing still eastward it gets narrower, so that when it reaches the Atlantic coast it is a mere fraction of what it was. In round numbers this immense region contains about 1,500,000 square miles, and its forest is made up of very few species of trees, the principal ones being pine, spruce, tamarack and aspen poplar.

Indeed eight species of trees may be said to constitute the whole arborescent flora of the region in question. The species are:—*Pinus Banksiana*, Lam., *Picea alba*, Link., *Picea nigra*, Link., *Larix Americana*, Michx., *Populus tremuloides*, Michx., *Populus balsamifera*, Linn., *Betula papyrifera*, Michx., and in less abundance and of more circumscribed range *Abies balsamea*, Marsh. On the south-eastern margin *Thuja occidentalis*, Linn., and *Betula lutea*, Mx., are occasionally met with but may be excluded when speaking generally. Willows of many species are found throughout the whole region but they seldom become trees.

Although the above trees occupy the area under discussion it must not be understood that they grow indiscriminately over the whole surface.

The tamarack or larch, as with us in the east, is still inclined to occupy the wet ground around muskegs, but as it nears its northern limit it leaves their vicinity and grows where the soil is drier and more heated in summer. The black spruce in the east prefers the boggy ground, but as it approaches its northern limit it seems to enjoy the drier ground and vies with the white spruce in occupying the last oases before the forest ceases altogether and the continuous barren grounds commence. Wherever the ground is sandy or rocky, or both, the Banksian pine flourishes, and as it passes from east to west it loses its low and scrubby character as is the case along the St. Lawrence and Lake Superior, though it is a much finer tree in the latter district, and becomes a handsome tree west of Lake Winnipeg. On the Beaver, the English, the Athabasca and the Clearwater rivers, between lat. 53° and 58°, it attains its greatest dimensions, and is there a stately tree over 100 feet high and having a diameter from 12 to 20 inches.

The four trees mentioned above are the conifers of the northern forest and may be classed as forming the sub-arctic forest proper. They keep their tree form to their utmost limit, not dwindling to mere shrubs as they do on mountain summits but forming outliers, in the barren grounds, of fairly developed trees even at their extreme limit. This being the case some other cause than the absence of heat must be given to account for this. From the statements of Mr. J. B. Tyrrell, who traversed the barren grounds last season, I am led to believe that the true reason for this barrenness is too much humidity in the air, and consequently a wet cold soil that scarcely rises a few degrees above freezing under the very best conditions, and in which trees could not exist, much less grow.

The poplars and birch grow under altogether different conditions from the conifers. The aspen in the east seems to be a poor sickly tree, very seldom having a thrifty look and preferring gravelly hillsides and borders of swamps. Its habit and appearance change wonderfully as we come upon it on the Canadian Pacific railway after passing out of the spruce and tamarack before reaching the prairie on our way to Winnipeg. Lying between the tamarack and spruce, and the prairie is the belt of aspen which is only a few miles wide along the railway but which extends from the international boundary in lat. 49° all around the prairie regions, and may be said to constitute nearly the whole forest growth of the prairies outside of the river valleys. North of the prairie it penetrates the coniferous forest wherever there is good dry soil, and is the bulk of the forest in the Peace River country and on the plains lying along the Liard and the Mackenzie. It may be said with truth that aspen forest means agricultural land wherever found, and as it is in southern Manitoba so is it on the Peace River plains and farther north. In the Riding and Porcupine Mountains and westward through the forests to Prince Albert and Edmonton, a distance of 800 miles, this species is found to be a fine tall tree. In many cases the bark is quite white and the round smooth trunk, rising from fifty to one hundred feet, with a diameter ranging from six to eighteen inches, is a remarkable object when seen in company with the brown barked gloomy looking spruce.

The aspen in its north-western home keeps out of the flood plain of the river valleys and never appears on islands or indeed on alluvium at any time. On the other hand, balsam poplar makes its home there and is seldom found anywhere else. On the Saskatchewan and all its branches the balsam grows to a large size, but these are but pigmies compared with those on the Peace, Athabasca, Liard, Slave and Mackenzie Rivers. On the islands in these rivers it grows to an immense size and it is no uncommon thing to see a tree over six feet in diameter without bark stranded on a bar. It is this species and the white spruce that are found as drift wood on the shores of the Arctic Sea, as they constitute the trees of the islands and flood plains of the Mackenzie and its tributaries, which are constantly changing and being reformed by the spring freshets. All the islands and points are constantly changing except when there is a jam of logs at their upper end. In many cases a few hundred yards walk will take a person from trees four feet in diameter to the lower end of an island where the young seedlings are just emerging from the mud. If the island or point be quite large spruce will take possession of the upper end before the wasting takes place, the old poplars will be smothered and rot, and the spruce will live on their remains. Spruce are never found on a new island.

The Canada balsam (*Abies balsamea*) and the paper birch (*Betula papyrifera*) are not very common and may be passed over with a few words. The birch is the more plentiful tree and has a wide range but is never a striking object or very plentiful. Besides using its bark for canoes, the Indians in the English River and Chipweyan districts make, in spring, a very

nice syrup from its juice, which before the advent of
" canned goods " served in place of the dried and canned
fruits now carried by travellers.

In another place I speak more in detail of the forests
of British Columbia and the Rocky Mountains, but a
few words may be necessary here to carry the sub-
arctic forests to the Pacific coast. The only known
change that takes place in the forest after reaching
the mountains north of lat. 53° is the substitution of
Pinus Murrayana for *Pinus Banksiana* and *Abies
subalpina* for *Abies balsamea*, which was left far to
the east. It may then be said that from lat. 53° west
to the Coast Range and the tundra of Alaska, with the
exceptions above stated, the same forest extends from
Labrador to within a few miles of the Pacific coast.

Crossing the summit of the Coast Range and descending towards the west, we meet with a different
forest composed chiefly of *Picea Sitchensis*, *Abies
amabilis*, *Thuya excelsa* and *Tsuga Mertensiana*, and
towards the south *Pseudotsuga Douglasii*, *Thuya
gigantea* and *Alnus rubra*. The moist winds from
the Pacific with the mildness of the winters combine
to produce on this coast a most exuberant growth of
every species, so that the forest is filled with a rank
vegetation and the stately trees stand rank behind
rank in serried phalanx forming a forest growth that is
unequalled in America, and extending from southern
Alaska to California.

NOVA SCOTIA AND NEW BRUNSWICK.

The forest floras of Nova Scotia and New Brunswick are practically identical and the climatic conditions are very similar in both provinces. On the side
towards the Gulf of St. Lawrence the same conditions
prevail as in Prince Edward Island, and the hardwood
timber is found much nearer sea level than along the
Atlantic coast and the Bay of Fundy. Northern New
Brunswick has a more continental climate and may be

compared with that of Quebec and northern Ontario. The following twenty-nine species, with the exception of *Tilia Americana, Juglans cinerea* and *Quercus macrocarpa*, occur in both provinces.

Tilia Americana, Linn. (Basswood).
Acer saccharinum, Wang. (Sugar maple).
    " *rubrum,* Linn. (Red maple).
    " *Pensylvanicum,* Linn. (Striped maple).
Prunus serotina, Ehrh. (Black cherry).
    "   *Pennsylvanica,* L. f. (Bird cherry).
Fraxinus sambucifolia, Lam. (Black ash).
    "   *Americana,* Linn. (White ash).
    "   *pubescens,* Lam. (Red ash).
Ulmus Americana, Linn. (Elm).
Juglans cinerea, Linn. (Butternut).
Betula alba, var. *populifolia,* Spach. (White birch).
    "   *papyrifera,* Marsh. (Canoe birch).
    "   *lenta,* Linn. (Cherry or black birch).
    "   *lutea,* Michx. f. (Yellow birch).
Quercus rubra, Linn. (Red oak).
    "   *macrocarpa,* Mx. (Mossy cup oak).
Fagus ferruginea, Ait. (Beech).
Ostrya Virginica, Willd. (Iron wood).
Salix nigra, Marsh. (Black willow).
Populus tremuloides, Michx. (Aspen).
    "   *balsamifera,* Linn. (Balsam poplar).
Pinus Banksiana, Lam. (Scrub pine).
    "   *Strobus,* Linn. (White pine).
    "   *resinosa,* Ait. (Red pine).
Picea alba, Link. (White spruce).
    "   *nigra,* Link. (Black spruce).
    "   *rubra,* Lam. (Red spruce).
Abies balsamea, Mill. (Balsam fir).
Tsuga Canadensis, Carr. (Hemlock).
Larix Americana, Michx. (Larch, tamarack).
Thuya occidentalis, Linn. (White cedar).

Owing to the influx of the cold winds from the Atlantic and the Bay of Fundy, the coast species are chiefly spruces and firs; but a few hundred feet of elevation above the river valleys bring us into a hardwood forest composed of maple, beech, ash and birch, with a sprinkling of spruce and pine, except in the western parts where spruce, fir and tamarack are the prevailing trees; in general terms this may be also said of Quebec, as the forests of northern New

Brunswick are almost identical with those of that province. The American elm is, as usual, found most highly developed in the river valleys, birch and red maple growing with it here as elsewhere in the eastern provinces.

A study of the conditions under which the forests of Nova Scotia grow and occupy the ground shows that the sea air is not congenial to the native hardwood trees except the birch. An examination of the trees of the inner slope of North Mountain near Annapolis shows that the conditions necessary for the growth of hardwood trees are those required for the full development of the apple, and it would be well for fruit growers to preserve with care the forests on the Bay of Fundy side of the beautiful Annapolis valley. Since the forests were cut away in the neighbourhood of Kentville, Wolfville and Grand Pré, the soil has become much wetter and in many places where formerly the soil did not require drainage it is now necessary. The cutting away of the forests and letting in of the sea air has allowed tamarack to grow where formerly beech and maple occupied the soil.

The tendency in Nova Scotia and New Brunswick is for the forest to re-clothe the soil, but when the hardwood trees of the original forests disappear, spruce, balsam, birch and tamarack take their place and everything shows that in that region the cutting away of the forests does not lessen the rainfall, but rather increases the deposition or brings the general air nearer to the point of saturation. The change in climate is causing a decline in grain-raising and increasing the area of drained soil devoted to fruit-growing and stock-farming.

In southern New Brunswick, *Juglans cinerea, Tilia Americana* and *Quercus macrocarpa* are found in some abundance, but they cannot be said to be common anywhere and they indicate a higher temperature as we pass from the conditions peculiar to the coast.

QUEBEC.

The forests of Quebec are still very valuable and very extensive and approach those of northern and central Ontario in the number and distribution of species. The conditions found on the New Brunswick border extend into Quebec and south of the St. Lawrence to Montreal. The same conditions obtain in the valley of the St. John River and up the Ottawa to its source. Except in the more southern districts, the elms, maples and beeches occupy restricted areas as they do farther east, but the general distribution is the same and the trees of Quebec with few exceptions are the trees of the Maritime Provinces. The following additional species enter Quebec but only along the Ottawa and St. Lawrence valleys.

Acer dasycarpum, Ehrh. (Broad-fruited maple).
Crataegus coccinea, Linn. (Red-fruited thorn).
Ulmus fulva, Michx. (Slippery elm).
        "   *racemosa*, Thomas. (Rock elm).
Celtis occidentalis, Linn. (Nettle tree).
Carya amara, Nutt. (Bitternut).
        "   *alba*, Nutt. (Shell-bark hickory).
Carpinus Caroliniana, Walt. (Blue beech).
Quercus alba, Linn. (White oak).
Populus monilifera, Ait. (Cotton-wood).
Juniperus Virginiana, Linn. (Red cedar).

None of the above trees are very abundant and the elms and bitternut are the only species that could be called common anywhere in Quebec. The hickory and nettle-tree cling to the St. Lawrence and are seldom seen elsewhere in the province.

The northern forests of Quebec are a part of the sub-arctic forests and are composed of only a few species of trees. The more valuable woods of commerce are found south of the watershed of the northern tributaries of the St. Lawrence and the Ottawa, and these constitute the present lumber regions of the province. Still farther south on both sides of the St. Lawrence and the lower Ottawa lie the fertile lands of the province that in the past had a mixed forest of hardwood trees where the ash, maple, birch, beech and elm gave character to the landscape and natural beauty to river, lake and shore. Many areas of mixed forest remain almost untouched in Quebec, and when these forests are cleared away hundreds of smiling farms will take their place. The two most important areas are the Lake St. John district, north of Quebec, and the very valuable and large tract of country towards the sources of the Ottawa.

ONTARIO.

Owing to the position and extent of Ontario its forests are not all of the same character and while in the north and north-west the species are identical with those found in Quebec, those in the south and south-western peninsula are quite distinct and may be said to be a reproduction of the northern Ohio and Pennsylvania forests. A few words will suffice for the north and north-west. What was said of Quebec north and south of the St. Lawrence watershed is applicable here. Only the species of the sub-arctic forest find a congenial home in this region and at the head of the streams flowing southward into the Ottawa and the great lakes are to be found the remnants of the noble forests that supplied material for the devastation of the last half century. It is truly appalling when the magnitude of the national interests at stake are considered, to view the spoilation which has been carried on quite recklessly under the protection of permits and licenses. When one is soberly told that this destruction was necessary in the interests of trade and for the development of the country, one is forced to deny the truth of such statements and to enter a protest against the fallacy concealed in them. If there had been any just or proportionate return to the state from such operations the objections might have less force, but when it is realized that for this splendid heritage the people of Canada have directly received only a nominal return in dues and bonuses, the responsibility for such a waste of resources, which should be guarded for the present and future generations of Canadians, is indeed grave. It is hardly a forcible argument to advance, that the money placed in circulation as wages to labourers employed in lumbering and the consequent local stimulation to trade or the enormous increase of private capital are a sufficient indirect gain. The cash paid as wages for such labour, labour which should have been used in the protection and development of these very forests, could never represent if multiplied many hundred times, the loss which has occurred owing to its misdirection : and the capital represents only a fraction of the use and value of the forests which should have been guarded for the public benefit. It is not yet too late to formulate a policy which will protect the sparse remains of this once dense forest and control them for the best interests of the whole country ; it is a policy which the present generation demands and the neglect of such a plain duty on the part of our legislators will only be an evidence of short-sightedness, of the triumph of party over patriotism, for which they will be visited with the just reprobation of those who will have to suffer from the present ill-considered action.

That part of the south-western peninsula of Ontario which lies west of Toronto has a flora quite distinct in many respects from any other part of Canada. Its position between lakes Ontario and Erie and along the latter lake accounts for this, and to this also is due its value as a fruit garden. The trees peculiar to this distant are :

Asimina triloba, Duval, (Cucumber-tree).
Liriodendron Tulipifera, Linn. (Tulip-tree).
Gymnocladus Canadensis, Lam. (Kentucky coffee-tree).
Cercis Canadensis, Linn. (Judas-tree).
Gleditschia triacanthos, Linn. (Honey locust).
Pirus coronaria, Linn. (Crab apple).

Crataegus Crus-galli, Linn. (Cock-spur thorn).
"   *tomentosa*, Linn. (Downy-leaved thorn).
Amelanchier Canadensis, T. & G. (June-berry).
Cornus florida, Linn. (Flowering dogwood).
Nyssa multiflora, Wang. (Sour gum).
Fraxinus quadrangulata, Michx. (Blue ash).
Sassafras officinale, Nees. (Sassafras).
Platanus occidentalis, Linn. (Button-wood).
Carya porcina, Nutt. (Hog-nut hickory).
"   *tomentosa*, Nutt. (White-heart hickory).
"   *microcarpa*, Nutt. (Small-fruited hickory).
Juglans nigra, Linn. (Black walnut).
Castanea sativa, Mill, var. *Americana*, Gray. (Chestnut).
Quercus bi-color, Willd. (Swamp white oak).
"   *coccinea*, Willd. (Scarlet oak).
"   *palustris*, Du Roi. (Swamp oak).
"   *tinctoria*, Bart. (Black oak).

In the above list there are twenty-three species
which represent a flora that has its affinities in the
south and gives an entirely different aspect to the
forests of the western peninsula when compared with
those of the east. One leading feature is the almost
total absence of coniferous trees and the great development of the hickories, the oaks, the button-wood,
the chestnut and the tulip-tree. The shrubs and herbaceous plants change with the forests, and scores of
species not found in other parts of Canada grow here
in profusion. The cucumber-tree was once common
around Niagara and Queenston, now it is so rare that
only the older people can tell one of its existence. In
June, 1892, I searched for days before I found a clump
fit to photograph. These were on the Niagara escarpment near Merritton. I have also found it fruiting
at Leamington, in Essex county. Although the sassafras is scattered through the old forest and is quite
a large tree, it is becoming scarce around clearings,
and is seldom planted. There are many fine specimens
about two or three miles from Niagara Falls on the
high road to Merritton and St Catharines. The Kentucky coffee-tree, honey-locust and Judas-tree are confined to Pelee Island and were not observed on any
part of the mainland except when cultivated, yet the

two former are quite hardy at Ottawa, and two fine specimens of the first species are now growing in front of Rideau hall.

Another peculiarity of the peninsula is that species which in other parts of the province are only large shrubs or very small trees, are here well developed, and have become fair-sized trees. Included in this group are four species of *Cratægus* and the June berry (*Amelanchier*), which in the vicinity of Niagara-on-the-Lake are very noticeable. Even the wild grape, *Vitis æstivalis*, has often a stem over four inches in diameter, and *Cornus alterniflora*, *Sambucus racemosa*, and *Viburnum Lentago* become trees, and in fence corners make a fine shade for cattle and sheep.

Were my paper intended to illustrate climatic conditions or the many lessons to be learned from the natural distribution of the forest, I might show from the wild grape, the plum, the wild apple and the wild cherry the economic importance of this district as a fruit producer. Only a few years since our own people believed that peaches and certain varieties of the grape could be grown only in favoured localities, yet the forest growth if read aright would have told them that with proper local shelter all the finer fruits of temperate climates were suited to the district under consideration, and not alone to this district but to the whole of Ontario along the St. Lawrence and Lake Ontario. With the exception of the peach, every other species can be profitably raised as far east as Ottawa, if proper shelter be forthcoming, for it is not a low temperature so much as unsuitable conditions that prevents the successful culture of fruits in Ontario. A lesson hard to learn is that shelter from nipping winds is just as necessary for vegetation as it is for the shorn lamb, and when horticulturists and others realize this to its full extent there will be fewer failures in fruit growing.

19

MANITOBA AND THE NORTH-WEST TERRITORIES.

The trees of the forests of this immense region are few in number and nearly all belong to the sub-arctic forest, and as a whole have been treated under that head. Two trees which we have had with us from Nova Scotia appear in Manitoba, but they are never found in much abundance and seldom out of the river valleys. These are the elm and the basswood. The green ash (*Fraxinus viridis*) and red ash (*Fraxinus racemosus*) are found in the valleys of the Red, Assiniboine and Souris Rivers but do not leave their valleys. On the other hand the over-cup oak (*Quercus macrocarpa*) forms thickets and open forests in many parts of Manitoba, becoming a fine tree at times, but dies out west of the Assiniboine above Fort Ellice. The elm disappears on the Red Deer River—not far west of Lake Winnipegoosis, and at its extreme limit is still a well-developed and large tree. The last sugar maple was left at McKay's Mountain, near Lake Superior, and the red or swamp maple disappeared at Rainy Lake, but a few basswood manage to reach nearly as far west as Brandon in the Assiniboine valley, and from thence westward all trees, apart from the species belonging to the sub-arctic forest, are of western origin, except *Populus monilifera* (cotton-wood) and *Negundo aceroides* (ash-leaved maple). These trees extend, in the river valleys, far out towards the Rocky Mountains, but do not reach them.

In the Cypress Hills west of long. 110° west, at an elevation of over 3,000 feet, the Rocky Mountain scrub pine (*Pinus Murrayana*) is found in abundance, and from this tree the hills take their name, the scrub pine of the east (*Pinus Banksiana*) being the cyprès of the French voyageurs. In the valleys of the rivers forming the South Saskatchewan two species of poplar (*Populus angustifolia* and *P. acuminata*) are found. These are a part of the more southern forest and are not known north of Medicine Hat.

ROCKY MOUNTAINS AND BRITISH COLUMBIA.

The trees of the Rocky Mountains may with few exceptions be classed with the western flora, and those that have not that origin belong to the sub-arctic forest, and have descended from the north along the mountains. The following list includes all the trees of the Rocky Mountains, a few of them occurring only on the western slopes facing the valley of the Columbia River.

ROCKY MOUNTAINS.

Populus tremuloides, Michx. (Aspen).
    "   *balsamifera*, Linn. (Balsam poplar).
Picea alba, Link. (White spruce).
    " *Engelmanni*, Engelm. (Engelmann's spruce).
Abies subalpina, Engelm. (Mountain balsam).
Pseudotsuga Douglasii, Carr. (Douglas fir).
Pinus flexilis, James. (Rocky Mountain pine).
    " *Murrayana*, Balfour. (Black pine).
    " *albicaulis*, Engelm. (White-barked pine).
Larix Lyallii, Parl. (Mountain larch).

Other species in the Columbia Valley and Selkirk Mountains.

Populus trichocarpa, Torr. & Gray.
Juniperus Virginiana, Linn. (Red cedar).
Thuya gigantea, Nutt. (Western white cedar).
Pinus monticola, Dougl. (Western white pine).
    " *ponderosa* var. *scopulorum*, Engelm. (Yellow pine).
Tsuga Pattoniana, Engelm. (Mountain hemlock).
    " *Mertensiana*, Carr. (Western hemlock).
Larix occidentalis, Nutt. (Western larch).

ADDITIONAL PACIFIC COAST SPECIES.

Acer circinatum, Pursh. (Vine maple).
    " *macrophyllum*, Pursh. (Broad-leaved maple).
Rhamnus Purshiana, DC. ("Barberry").
Prunus emarginata, Walp. (Western bird cherry).
Pirus rivularis, Dougl. (Western crab apple).
Cornus Nuttallii, Aud. (Western flowering dogwood).
Arbutus Menziesii, Pursh. (Madrona).
Salix Scouleriana, Barratt. (Western willow).
Thuya excelsa, Bong. (Yellow cypress).
Taxus brevifolia, Nutt. (Western yew).
Pinus contorta, Dougl. (Western scrub pine).
Picea Sitchensis, Carr. (Menzies spruce).
Abies grandis, Lindley. (White fir).
    " *amabilis*, Forbes. (Mountain fir).

The bulk of the forest in the Rocky Mountains south of lat. 53 is made up of white spruce, Engelmann's spruce, black pine, Douglas fir and balsam fir. These five species include at least 90 per cent. of the forest growth, the remaining 10 per cent being made up of the other five species. Of these *Pinus flexilis* is found only on the margins of the rivers issuing from the mountains, and the poplars in the valleys and open spaces where the original forest has been burnt off. On the other hand *Pinus albicaulis* and *Larix Lyallii* form a zone more or less pronounced at the extreme limit of trees, about 7,000 to 7,500 feet altitude, and in September the latter tree stands out very distinctly owing to the changing of its leaves from green to yellow.

All the valleys are filled with white spruce, and the mountain slopes, where gravel or sand predominates, are covered with pine. As we ascend above 5,000 feet, the pines are left behind and spruce and fir with Douglas fir take their place.

Descending from the Rocky Mountain summit by the Kicking Horse Pass, we meet the western cedar as a mere shrub, but in the Columbia valley it becomes a gigantic tree, often having a diameter of ten feet, in the valley of Beaver Creek. Ascending the slope on the west side of the valley we come at once into a belt of the western hemlock and white pine, which is characteristic of all the mountains from here to the Coast Range. Above these trees, but often intermixed with them, as at the Glacier hotel, Selkirk Mountains, Patton's hemlock is found capping the mountains or forming the last groves on their sides. On the Coast Range a change takes place, and the upper slopes are clothed with this tree and the white fir (*Abies amabilis*). Fine groves of this shapely tree are to be seen here, and the difference between it and the Rocky Mountain species (*Abies subalpina*) is very apparent, as the former has green cones and the latter bright purple ones. Descending the Columbia River, groves of the western larch are seen below the Upper Arrow Lake, and this fine tree is not uncommon on the lower slopes of the mountains on both the east and west sides of the Gold Range.

Generally speaking, all the valleys throughout both the Gold and Selkirk ranges are filled with cedar and spruce, and the mountain slopes are covered with Douglas fir and hemlock. The trees are in all cases well-developed, and from their size are suited for any purpose. This is the character of all the timber from the Columbia valley to the western slopes of the Gold Range. The valleys of the streams discharging westward from the latter range into the Eagle and Spullamacheen Rivers and Shuswap Lake are also filled with fine timber of the same species. Passing westward from these mountains we come gradually into a drier region, and the country becomes open, with only scattered groves or single trees on the lower slopes and plateaus, and the yellow pine (*Pinus ponderosa*) so characteristic of the dry interior of British Columbia is the chief feature in the landscape.

The light rainfall east of the Coast Range in British Columbia prevents the growth of a continuous forest outside the flood-plains of the rivers so that yellow pine and Douglas fir are scattered over the Okanagan and Kamloops country until we reach an altitude of about 3,500 feet. Above this is a belt of dense forest composed chiefly of spruce and black pine (*Pinus Murrayana*) with which is mixed, in places, a considerable quantity of Douglas fir. This forms a zone of from 2,000 to 3,000 feet above which the forest thins out and grassy meadows, with beautiful groves of fir, cap the mountains.

The transition from the arid region of British Columbia to the humid coast district is a sudden one. As soon as the summit of the range is passed a change occurs, and descending by the valley of the Fraser, this is noted a few miles above Boston Bar where the mountain barrier closes the valley to the moisture-laden winds from the Pacific. Descending into the lower valley of the Fraser causes little change in the trees outside the flood-plains, but they at once increase in size and more than double their height. It is in the lower Fraser valley that we first see the Pacific coast forest and are lost in wonder at the height of the Douglas fir, Menzies spruce and the western cedar. Trees of Douglas fir 300 feet high and ten or twelve feet in diameter were formerly common, and many fine specimens still remain. A visit to Stanley Park at Vancouver will satisfy the most skeptical, and the remnants of the former forests seen there give full assurance that neither the size nor the number of trees in the old forests was exaggerated. The samples seen on this peninsula between New Westminster and Vancouver city will exemplify the forests of some parts of Vancouver Island and the coves and deep inlets of the mainland.

The broad-leaved maple is a coast species, but ascends the Fraser almost as far as its junction with the Thompson, and before it disappears dwindles to little more than a shrub. The arbutus is seldom seen on the mainland except on rocky points jutting into the sea, but it ascends the north arm of Burrard Inlet for a mile or two. From Burrard Inlet northward, the coast forests of the mainland change gradually, so that the sequence of trees on a mountain near Vancouver city will illustrate the gradual change on the coast, with one exception—Menzies spruce. This species is a very fine tree on Burrard Inlet and continues so, far into Alaska, while the Douglas fir seems to be at its best here, and begins to diminish in size and numbers towards the north end of Vancouver Island. It gradually becomes intermixed with hemlock (*Tsuga Mertensiana*) and yellow cedar (*Thuya excelsa*) to the north and eventually disappears, and the coast forests are then composed of spruce, hemlock and yellow cedar only.

The species on the mountain summits of the mainland are little known, but reasoning from what we know of Vancouver Island we can safely say that *Tsuga Pattoniniana* and *Abies amabilis* are the principal trees. These are intermixed on the upper slopes with *Thuya excelsa* and *Tsuga Mertensiana*, while on the middle slopes *Pinus monticola* is well developed.

VANCOUVER ISLAND.

There are no trees on Vancouver Island that are peculiar to it, except a doubtful poplar and only one which is not found on the mainland—the western white oak (*Quercus Garrayana*). This tree covers a considerable area of rocky ground around Victoria, and is found at Departure Bay and in some quantity at Comox, but in the latter locality it is of little value. Douglas fir is the chief forest tree throughout Vancouver Island. On the south it is mixed with giant cedar and balsam fir. On the mountain slopes this tree with white pine, yellow cedar and hemlock constitute the forest, and at an altitude of 5,500 feet it holds its own with *Abies amabilis* and *Tsuga Pattoniana*. As we pass to the north the forest changes and the mountain trees descend so that the yellow cedar, first seen on Mount Benson, near Nanaimo, at an elevation of 1,200 feet, reaches the coast some distance south of the north end of the island.

The trees which give character to the Vancouver Island vegetation are the arbutus, flowering dogwood and broad-leaved maple. The former with its large laurel-like evergreen leaves is a living proof of the mildness of the climate, and its red inner bark and green leaves as it is seen standing on a rocky point or jutting rock along the coast relieves the sombre aspect of the thick forests of Douglas fir. The dogwood may often be seen in company with it, its white involucre, over three inches across, covering the tree with a mantle of white, broken here and there by protruding leaves.

In conclusion I may say that including Vancouver Island a coniferous forest may be said to extend from the Pacific to the Atlantic, bounded on the north by the tundra of Alaska and the Barren Grounds of the Dominion, and southerly with a varying border until it meets and intermingles with the poplar forests of the North-West Territories. Passing still eastward the poplar mixes with it to the south until after passing Lake Superior it gradually merges into the deciduous forests of Ontario, southern Quebec and the elevated and interior region of New Brunswick, Nova Scotia and Prince Edward Island.

After all has been said about our waste both by myself and others it is evident that we have woodland enough in the north to supply every demand that may be made upon it for many generations, but like everything that is valuable, it is hard to get at. When it will be wanted none can say, but that it is there in incalculable quantities is absolutely certain. A belt 200 miles deep and 3,000 miles wide gives us an area of 600,000 square miles, but we are quite safe in estimating it at 1,000,000. The poplar forest and the mixed growth to the north of it extends from Edmonton to Winnipeg, a distance of about 900 miles, and averages over 50 miles in width, which gives an area of 45,000 square miles of aspen forest for the use of the settlers who will by degrees occupy this region, for the aspen districts have, as a rule, good soil.

CHAPTER III.—APPENDIX.

The Lumber Industry of Canada.

By A. H. CAMPBELL, Jr., B.A., TORONTO.

NEXT to agriculture, the products of the forest constitute by far the most important industry of the
Dominion from an economic point of view. The last
census reports show that over $100,000 000 of capital
are invested in industries directly dependent on the
forests for their raw material; paying in wages alone
over $30,000,000, and turning out an annual product
of over $125 000,000.

There are few questions of more vital importance to
Canada than the future possibilities of the forests, but
reliable information on the subject, is almost unattain-
able. The concensus of opinion is, however, that under
present conditions, as far as the white pine—the most
important wood of the country—is concerned, the end
is within measurable distance, the destruction of it
being enormous, while its growth to maturity is exceedingly slow. The spruce of the Maritime Provinces
should last for a long time; and the extensive forests
of British Columbia are almost untouched; but with
these two exceptions, it will not be long before Canada
ceases to be a wood-exporting country, unless some
proper system of forestry is introduced.

The important points which will have to be considered, are a proper system of reproduction by
judicious cutting and re-planting, and the preservation
of the forests from fire. There are many and great

difficulties in the way owing to the enormous areas to
be protected, but the various Provincial Governments
are beginning to realize how important the matter is,
and initial steps have been taken to secure the desired
end.*

The two most active agents in the destruction of
the forests, are the lumberman and fires; and of
these two, the fires are harder to combat and do
more damage, as trees of all kinds are killed, and
the younger ones are thus prevented from ever reaching maturity. Forest fires are usually the result of
carelessness on the part of some inexperienced person,
and when once started during the dry months of the
summer, they sweep through the forests at an incredible rate, and are absolutely uncontrollable till extin-
guished by heavy rains. The pine forests are more
subject to fires than the spruce, as in the former there
are more tops and branches left in the woods after the
merchantable timber has been taken out, while in the
spruce forests, the growth is more dense and close to
the ground, which keeps the moss and underbrush
damp and prevents fires from spreading so easily. In
almost all the provinces, very stringent laws have been
passed to prevent, as far as possible, fires in the forests,
and these, supplemented by a body of fire rangers,
have had a very beneficial effect, and fires are less frequent than formerly. The effect of a fire is not so
injurious to the large trees as might be supposed. The
timber itself is seldom destroyed; but all the trees
large and small, which have been scorched by the fire,
are killed, and almost immediately the eggs of the
pine-borer or sawyer, are deposited in the bark. These
eggs are soon hatched into a grub, which in time
bores into the tree in every direction, spoiling it for any
but the commonest uses. Fires invariably occur in the

* Since the above was written a Forestry Commission has been
appointed by the Ontario Government, to enquire into the best
means of conserving the timber of the province.

dry months of the summer, and if the timber so scorched is cut during the following winter, little appreciable damage is done; the drawback being the necessity of cutting it to save it from loss, whether it is wanted or not. Settlers are not encouraged to go on to lands covered with uncut timber, as they increase the danger of fire, but after the timber has been cut, the land, if suitable for cultivation, is available for settlement.

As a rule the cutting of the small and growing timber is now prohibited by the Crown; and if forests—particularly the spruce forests—on which the young timber is spared are protected from fire, they will in a short time reproduce themselves.

From a commercial point of view, the white or Weymouth pine is by far the most important tree of the country. The next in order is the spruce, which grows in more northern latitudes than the pine, and is the principal timber of the Maritime Provinces. British Columbia has still large areas of virgin forests in which the Douglas fir, spruce, cedar, and hemlock, are the most common trees; but owing to the want of a market in the west, the trade has not yet been developed to any extent.

No account is here taken of hardwood timber, for as it will not float in the streams, it is only available in the immediate neighbourhood of the railroads or navigation; and the cutting of it is confined to the settled parts of the country.

At one time a large quantity of timber in the rough, was exported to the United Kingdom, where it was manufactured into boards, etc., but this trade has been much curtailed in recent years, and the timber is now usually manufactured at the saw-mills in Canada, and then shipped to the different markets.

Within the last few years a large quantity of wood-fibre, or wood pulp has been made from spruce, and this promises to become a very important industry of the Dominion.

The timber rights in almost all parts of the Dominion are vested in the several Provincial Governments, and when disposing of these privileges the Crown retains a large interest in the timber, only selling a right to cut, commonly called a timber license. These timber licenses are offered for sale at public auction, and entitle the licenciate to cut all kinds of timber off the lands so licensed, on payment of a small annual ground rent, and a royalty on the timber as it is cut. The form of licenses varies in the different provinces, some being in perpetuity and others for a fixed period, usually twenty-five years. The value of a timber license depends on so many factors,—quantity and quality of timber, cost of operating, etc.--that only the most experienced men. can with any certainity arrive at an accurate idea of it.

The working operations of the lumberman are of a very varied nature and from one and a half to two years usually elapse before the standing tree is cut down, manufactured and put on the market, and on few things is the success of the industry more dependent than the climate and season of the year.

In September a small force of men is sent into the woods to put up the necessary buildings for the winter's work. After a location, convenient to the work and near fresh water, has been found, the camp or shanty is built. It usually consists of two large buildings for the men—one for them to sleep in, and the other for the kitchen and eating room,—a stable and barn for the horses and provender, a store house for supplies, a smithy, an office where the foreman and clerk sleep and where the business of the camp is conducted. These buildings are made of logs hewn to a square, the joints being filled up with clay, and are warm and comfortable. The food while plain is always good, and as the outdoor life is a very healthy one, it has its attractions for the men. The fascination about life in the bush is akin to that exercised by the sea.

A gang for a camp usually consists of forty men and four teams, with ten to fifteen additional teams when the logs are being hauled to the stream. Provisioning a camp of this size for from six to seven months is necessarily a very important matter, as all the supplies have to be brought in from the outside world. Frequently they are drawn in on sleighs over the snow during the preceding winter, but the rivers often afford a way of getting them in just before they are required. The shanties offer an excellent market to the farmers for their produce, and also provide work for the men and horses during the winter when there is little to be done on the farm.

The work proper of the camp is commenced by the choppers, who fell the trees and cut them up into proper lengths. An experienced chopper can bring a tree down in the exact direction required, unless one side is very much overloaded with branches, and there are few more impressive sights than the fall of these giants of the forest as they come crashing down. The axe is used for felling the tree and the saw for cutting it up into proper lengths—or saw-logs—after it is down. The length of the logs are twelve feet and upwards, according to the uses for which they are required. After the tree is cut up, the logs are dragged one at a time by a team and piled in a heap called a "skidway" the front of which is on one of the roads leading to the river. During this operation which is called "skidding" the work for the horses is not very heavy but is very trying and at times dangerous. The ground is rough and the horses have to find footing where they can, but after a little practice and with experienced teamsters the horses become accustomed to the work, and will go through the woods over fallen trees and other obstructions without hurting themselves. While being put on the skidways each log is measured by a Government Inspector, who calculates its cubical contents, and on this measurement the royalty due to the Crown is paid. The distinguishing brand of the owner is also stamped on the logs, so that they can be identified at any time. The work of chopping and skidding goes on till the snow commences to fall which is usually about Christmas. While this is going on, another gang of men are at work making roads from the skidways to the rivers. These roads have to be very carefully laid out, the underbrush, etc., is then cleaned away, holes filled up and good grades made. The frost and snow complete the work, making excellent roads in what would otherwise be an impassable country. Extra teams are then brought in, and the logs are hauled on sleighs from the skidways to the banks of the streams, where they are rolled into the water. Frequently the roads are cleaned off every night with a snow plough and then sprinkled with water, so that by morning there is a smooth surface of ice all the way. The teams are then able to haul enormous loads, as there is usually a gentle slope to the river.

As soon as the first signs of spring appear a gang of men, called river-drivers, is organized to bring the logs down the rivers. The melted snow and spring rains make a great flood of water on all the streams, and this is taken advantage of by the lumberman to float his timber down many creeks which in summer are dry. This natural flood is artificially increased and regulated by dams erected at the lower ends of lakes where the water is held, and gradually let off when wanted, by means of sluice gates in the dams. Slides or chutes in the shape of a large trough, made of timber, through which part of the stream is diverted, have often to be constructed to take the logs over high falls or very rough rapids where they would otherwise be damaged by the rocks. When the logs have to be taken across a lake they are enclosed in a boom consisting of long pieces of timber chained together at the ends, and the raft is then towed across by tugs or a capstan. A boat called an alligator, from its amphibious character, is often used. It is built with a flat bottom shod with steel, and is equipped with paddle wheels and a powerful windlass driven by an engine. It is sent ahead of the raft paying out the tow line and after being fastened to the shore or anchored, the tow line is wound in on the windlass dragging the raft up to it, when the operation is repeated. When a rapid or fall has to be passed by the boat, a road is cut out along the shore, the tow line taken ahead and made fast, and the boat pulls itself across the portage by means of the windlass. In their work the river-drivers have to be constantly on the logs, and the ease with which they can cross the stream or go down a rapid on a round log is wonderful. Their boots are shod with sharp spikes to keep them from slipping, and they carry a pole with a pike on the end, with which they keep the logs moving, and which also acts as a balancing pole. Considering the risks these men run, and the large proportion engaged in the work who cannot swim, accidents are comparatively rare. The logs frequently get jammed across the river and in the rapids as well as stranded along the shores, and the river-drivers have to keep working at them continually to get them all safely down, the current in the streams, and the wind, being of great assistance to them.

The driver's work is at an end when the logs have been brought down to the saw-mills. The mills are as close as possible to the streams from which the timber comes, and at some point, whence the manufactured article—or lumber as it is called—can be cheaply forwarded to the markets. They are generally driven by steam power, the sawdust and other waste necessary in the process of manufacturing, providing ample and cheap fuel on the spot. The machinery used for cutting up the logs is necessarily of a very heavy and complicated kind. The circular or rotary was for a long time the only known saw, but in recent years has been replaced by large band-saws which save a great deal of timber by a finer saw-cut. They are made of a thin sheet of endless steel about ten inches wide, one-sixteenth of an inch thick, and usually about fifty feet long. The saw is hung like a belt, on two large wheels one above the other from which it gets its motion. The log is on a carriage which moves it past the saw, one board being cut off each time the log is taken up to the saw. Upright or gang saws are also used, consisting of a number of saws hung in a frame working perpendicularly, and through which the log is fed by means of rollers, the whole log being cut up into boards at the same time. After the boards come from these saws the edges and ends are squared off by other machines and the boards are then piled in an open pile to season. The drying process takes from four to eight weeks according to the time of the year and the quality of the lumber. The product is then ready to be forwarded to the markets. As lumber is a very bulky product shipments are more largely carried on by boats than by railways, and the magnificent waterways afforded by the Great Lakes and canals play a very important part in this industry.

The export trade is at present about equally divided between shipments to Great Britain and the United States, the figures for the year ending 30th June, 1896, being

Exports to Great Britain........$12,187,000.
Exports to United States....... 13,528,000.

The mills on the Ottawa and St. Lawrence Rivers and those in the Maritime Provinces as a rule ship to Great Britain, while those on the Upper Lakes find a market in the United States. The stock for the English market is usually cut three inches thick, and on arrival it is re-cut to the required sizes. The standard thickness for the United States market is one inch. If the high import duties on lumber proposed by the United States should become law it will probably have the effect of diverting shipments from that country to Europe.

The business of the lumberman is of such a complex kind, and the fact that operations are carried on in so many different places, make it necessary to exercise the greatest care and economy in every department to insure a successful result. There are also a great many contingencies over which the lumberman has no control, he being almost more dependent on the weather and climate for succeeding in his work than even the farmer. In many cases great fortunes have been made in the business, but taking it as a whole the profits made are small when the risks incurred are considered, and it is only after long—and frequently dearly bought—experience, that success can be hoped for.

CHAPTER IV.

The Mineral Resources of Canada.

By A. P. Coleman, M.A., Ph.D., Professor of Mineralogy in the School of Practical Science, Toronto.

FOR a country possessing almost half of North America, Canada has been surprisingly slow in developing its mineral resources, so much so that many foreigners and not a few Canadians have doubted whether Canada would ever become of importance as a mining country; but the last few years have shown so decided an advance that Canadians are beginning to hope that their half of the continent will prove as widely metaliferous and as rich as the great republic to the south.

The slow growth of mining in Canada is to be attributed, not to a lack of important mineral resources, but to a variety of circumstances connected with the settlement of the country, its areal geology and its climate. A population of about 5,000,000 in an area of more than 3,400,000 square miles of territory represents an average of only one and a-half individuals per square mile, and since the two most populous provinces, Ontario and Quebec, including more than two-thirds of the whole number of inhabitants, were established mainly as farming communities on the fertile palæozoic border of the great archæan shield where no deposits of ore or coal can be looked for, the bulk of the population have grown up with no knowledge of mines and with the thrifty virtues of farmers and merchants averse to risking their savings in an untried and hazardous occupation.

Almost all the provinces and territories of the

Dominion possess mineral resources of importance, but only two have been large producers as mining regions, Nova Scotia on the Atlantic coast and British Columbia on the Pacific; and it is remarkable that both these provinces have been producers chiefly of gold and of coal, although the eastern deposits are of an entirely different character from those of the vast and mountainous western province.

In studying the mineral wealth of Canada the most important sources of information are the reports of the Geological Survey and of the mining departments of Nova Scotia, Quebec, Ontario and British Columbia; but in the following short summary of our mineral development it will be unwise to give very detailed references to the authorities consulted. It is proposed to deal first with the more important metals and then with non-metallic minerals.

METALS OF CANADA.

Gold. Gold has been found in most of the provinces and territories of Canada, but has been mined to any serious extent only in Nova Scotia and British Columbia. The gold fields of Nova Scotia are in quartzites and slates of Cambrian age, the auriferous quartz occurring as saddle-shaped beds at anticlines or as small veins cutting these saddles, the interbedded quartz being low grade but in large quantity, while the fissure veins though usually very narrow are often rich. The relationships mentioned are much like those found at the Bendigo gold field in Australia. Hitherto the mining done has been chiefly on the narrow rich veins, but attention is now being directed, often with much success, to larger bodies of low grade ore. Owing to its favourable situation, healthful climate and plentiful supply of good labour, gold mining can be carried on in a most economical way, in some instances the whole cost of mining and milling the ore being not more than $1.65 per ton.

Nova Scotia is not a new mining country, gold having been produced in considerable quantities for more than thirty-five years, always by quartz mining, no placers of importance having been discovered. Since 1861 the annual product has averaged about $350,000, and the whole amount of gold produced up to the present is considerably over $12,000,000. That lower grade ores are now being treated than formerly is shown by the fact that in 1893, as reported by the Geological Survey, the average amount of gold per ton of quartz was $8.68 (a pennyweight is approximately worth a dollar), while in former years the quartz seldom ran below $12 per ton.* In 1896, 25,596 ounces were produced from 65,873 tons of ore.†

In the Province of Quebec auriferous gravels have for more than sixty years been known to exist on the Chaudiere River and its tributaries, but comparatively little work has been done on these placers, the whole quantity of gold obtained amounting, it is said, to only about $2,000,000. Interest in these old gravels, which frequently carry coarse gold, is being revived, however, and they may yet prove of importance. In 1893, a product of 872 ounces is reported by the Geological Survey and almost twice as much in the following year.‡

Gold was first found in the Province of Ontario in 1866, but in the thirty years that have followed, comparatively little gold has been produced until within the past year or two. There is now, however, every prospect that its output will rapidly increase. Unlike the adjoining Province of Quebec, Ontario contains no placers, the whole of the gold being obtained from quartz veins found at or near the contact of bands of huronian schist with areas of Laurentian gneiss or

---

*Geol. Sur. Can., vol. vii. 113 S. and 115 S.
†Rep. Dept. Mines, Nova Scotia, 1896, p. 30.
‡Ibid., 106 S.

granite or eruptive bosses of the latter rock. Gold has been found at various points for nearly a thousand miles along almost the whole length of the northern Archæan portion of the province, but the richest and most numerous finds have been made in a region two hundred and fifty miles long and about half as broad, lying to the west of Lake Superior. The returns of gold for the province are somewhat less than $33,000 for the year 1893, and the same for the following year. In 1895, $50,281 were produced, while the sum for eleven months of 1896 was $142,605, an average of $14.83 per ton of ore crushed, whereof $12.30 or 83 per cent. of the whole was free milling.*

There is every prospect that a number of mines will be producing gold in 1897, and that the total will rapidly increase. The area of auriferous country is so enormous and the ores as a whole so easily treated, being like those of Nova Scotia, mainly free milling, that within a few years a very large output may be expected.

Manitoba contains part of the gold bearing Huronian rocks, but up to the present no gold has been mined in the province.

A small quantity of gold is washed from the bars of the Saskatchewan River near Edmonton in the western territories; and a much larger amount from the Yukon region in the far north-west adjoining the Alaskan gold region, the total amount being estimated at $140,000 in 1894.†

British Columbia has long been known as a gold mining country, and may be looked on as the north-westward continuation of the great auriferous and argentiferous belt of the Western States from which so many hundreds of millions of dollars worth of the precious metals have been obtained. The best account

*Bulletin No. 1, Bureau of Mines of Ontario.
†Geol. Sur. Can., vol. vii., p. 106 S.

of gold mining in the province is to be found in Dr. G. M. Dawson's " Mineral Wealth of British Columbia," from which the present brief statement is mainly derived. Until within a few years all the gold of the province has been obtained from placer deposits, but of late quartz mining is gaining ground and before long will probably become much the more important of the two.

Gold mining on any extensive scale began in 1857, and the following year a rush of miners entered the province from the partly exhausted placer grounds of California, and worked many bars on the Fraser and its tributaries. In 1860, $2,228,543 were obtained, and in 1863 the high water mark of $3,913,563 was reached, the mines of Cariboo, in the northern part of the province furnishing a large part of the sum. For five years the annual product exceeded $2,000,000, and afterwards remained above $1,000,000 until 1881. From this time there was a rapid decline, until in 1893 the amount was only $379,535, a sum almost equalled by the little Province of Nova Scotia.

About this time the effects of hydraulic mining on the large scale began to make themselves felt, and a few quartz mines began to operate. Since then the Rossland region on Trail Creek near the American boundary has suddenly sprung into notice with its great deposits of pyrrhotite and chalcopyrite in rocks of a dioritic character. These ores are not generally free milling but are often rich enough to give good returns when smelted. In 1895 the gold output of the province was estimated at $1,290,545, and in 1896 at $1,788,206, so that the prospects for the future are very favourable. The whole amount of gold obtained in British Columbia up to 1896 is estimated in the report of the Minister of Mines of the province to be $57,704,855.*

---

*An. Rep. Minister of Mines, B.C., for 1896, pp. 497-8.

Platinum. The occurrence of platinum in the Chaudiere gold placers of Quebec was noted many years ago, but the amount obtained was insignificant. During the past ten years a small quantity has been furnished by the placers of British Columbia, particularly in the Similkameen region, the annual value as reported by the Geological Survey ranging from $1,000 to $10,000. A small amount of platinum is found associated with the gold and nickel ores near Sudbury, as the arsenide, sperrylite, hitherto known only in that region.

Silver. Three of the provinces have produced silver in larger or smaller quantities, Quebec, Ontario and British Columbia. The amount coming from Quebec ranges from 101,318 to 191,910 ounces per annum, but no silver ores proper are mined, the metal being obtained only from the cupriferous pyrites of the Capelton district of the Eastern Townships.[*]

In Ontario the fate of silver mining has been very uneven, and its history is almost romantic. The most important mine is that of Silver Islet near the mouth of Thunder Bay on Lake Superior. This little reef of rock was first worked in earnest in 1870, a shaft being sunk ultimately to a depth of 1,230 feet. The ore was at times immensely rich in native silver, argentite and other argentiferous minerals, and the total production up to 1884 amounted to $3,250,000. In that year the mine filled with water owing to the failure of the coal supply for the pumping engines, and has never been worked since. An excellent account of this mine is given by Mr. Blue, Director of the Bureau of Mines, in his report for 1896.

A number of silver mines showing similar rich ores have been worked on the mainland a short distance west of Thunder Bay in Cambrian (Animikie) slate overlain by sheets of diabase ; but none of them have

---

[*]Geol. Sur. Can., vol. vii., p. 118 S.

approached Silver Islet in productiveness. In 1891 these mines marketed 225,633 ounces of the metal, but owing to the rapid fall in the price of silver since that time they have all ceased work for several years. There is, however, some prospect that one or two of them will commence work again shortly. Probably the whole silver product of the Thunder Bay district does not exceed $4,300,000.[*]

British Columbia until recently produced only a few thousand ounces of silver per year; but the discovery of rich silver lead ores in the Kootenay region has rapidly increased the yield of the province. The argentiferous galenas and chalcopyrites occur in slates and schists, partly of Archæan age, partly much later, and in intrusive granites, the ores being entirely different in character from those of Thunder Bay. There are three mining districts producing silver lead or silver copper ores in West Kootenay, Slocan, Nelson and Ainsworth, the Slocan district having much the largest output. In 1894, according to the report of Mr. W. A. Carlyle, Provincial Mineralogist, West Kootenay had a product of 732,630 ounces of silver; in 1895 of 1,434,368 ounces; and in 1896 of 3,106,601 ounces, valued at $1,983,000, showing a very rapid increase. Owing to the small number of smelters in that part of the province most of the ores are shipped to points in the States to the south; but it may be expected that before long the expense of a long transport will be avoided by erecting smelters sufficient to treat all the ores of the region.

Lead. British Columbia is the only province of the Dominion mining any appreciable amount of lead and in this case the galena is mined really as an ore of silver in which the lead may be looked on as an important by product. On this account the output of lead advances *pari passu* with that of silver. The lead product comes from the Slocan and Ainsworth

---

[*]Geol. Sur. Can., vol. vii., p. 118 S.

districts of West Kootenay chiefly, the Nelson district
affording only silver copper ores. Mr. Carlyle reports
10,681 tons from West Kootenay in 1896, valued at
$636,579. As the silver from these two districts is
reported at 2,328,367 ounces for that year, each ton of
lead (of 2,000 lbs.) appears to have contained on the
average 218 ounces of silver.

Copper. Three of the provinces, Quebec, Ontario
and British Columbia, are somewhat important producers of copper, but rather as a by product than as
the primary object in mining ; and none of the metal
is refined in Canada, all being exported in the form of
ore or of matte to the United States. In the Capelton
district of the Eastern Townships, Quebec, the copper
occurs in small amounts in the pyrites mined chiefly
for the production of sulphuric acid. The Geological
Survey reports that in 1894, 2,176,430 lbs. were produced in this district, the value being placed at
$206,761.*

The copper production of Ontario may be divided
into two periods, an earlier one between 1846 and 1876
when a number of copper mines were worked on Lake
Huron, the total product being valued at $3,500,000 ;
and a later period since 1885 when the nickel copper
mines of the Sudbury district came into operation. The
matte resulting from the smelting of these ores contains nickel and copper, in nearly equal proportions.
In 1895, 2,365½ tons of copper were contained in the
matte, chiefly the product of mines belonging to the
Canada Copper Company.†

British Columbia's copper output is of very recent
date, since it comes mainly from the new Nelson mining district in West Kootenay where operations have
been carried on for only three or four years, and
the amount mined is rapidly increasing. The real

---

*Geol. Sur. Can., vol. vii., p. 47 S.
†Bur. Mines, Ont., 1895, p. 11.

value of the ore lies, however, in the gold and silver it
contains. In 1896, the amount coming from Nelson
is reported at 2,237,921 lbs., valued at $111,896.*
Several other mines are opening up in different parts
of the province and the production may be expected
to advance greatly within the next few years. The
output of copper for the whole Dominion has averaged
between seven and eight million pounds during the last
five years.†

Nickel. In the production of the fine metal nickel
the Province of Ontario has but one serious rival in the
world, the French penal colony of New Caledonia,
whose mines at present furnish a somewhat less amount
than the Sudbury district. The nickel ores of Ontario
are very like the gold ores of Rossland in British
Columbia, consisting of a mixture of pyrrhotite (magnetic pyrites) and copper pyrites. These sulphides
form enormous masses near the margin of large areas
of diorite, or rather of weathered gabbro of Huronian
age, the nickel contents of the ore averaging from 2½
to 10 per cent., the lower average being the more
common. It is worthy of note that pyrrhotite from
other parts of the country found in association with
Laurentian rocks is almost barren of nickel. The
amount of ore available in the Sudbury district is,
however, sufficient to supply the consumption of the
world for an unlimited time to come. Nickel mining
began at Sudbury in 1889 and reached an output of
4,626,627 lbs. in 1891, but the production has somewhat
fallen off in later years, though Mr. Blue, Director of
the Ontario Bureau of Mines, reports 4,631,500 lbs. in
1895, a slight increase over the figures for 1891. The
properties of steel are so greatly improved by an addition of two or three per cent. of nickel both for

*Bur. Mines, Victoria, B.C., Bull. No. 3, p. 38.
†Geol. Sur. Can., vol. vii., p. 47 S., and Rothwell's Mineral
Industry, vol. iv., p. 627.

structural purposes and in the manufacture of armour that it is probable the consumption of this metal will greatly increase in coming years, especially as the price is falling through improved methods of refining it. In 1890 it was quoted at 65 cents per lb., at present the quotation is 35 to 36 cents per lb.*

Iron. All the provinces of the Dominion possess more or less important deposits of iron ore but at present very little mining or smelting is going on except in Nova Scotia and Quebec. The former province mined 102,201 tons of iron ore in 1893, but the amount fell off to 89,379 tons in the following year. Three blast furnaces, two using coke for fuel and one charcoal, were in operation in 1894. In Quebec, 22,076 tons of bog and lake ore were mined in 1893, and two charcoal furnaces are in operation producing an excellent quality of cast iron. A single coke furnace has been in blast for some time near the city of Hamilton, Ontario. The whole product of iron in the Dominion in 1894 was 49,967 tons worth $646,447. A portion of the Nova Scotian pig iron was converted into steel.†

Formerly a considerable quantity of magnetite and hematite was exported from Quebec and Ontario to the United States, but the imposition of a heavy duty and the advent of cheap ores from the States south of Lake Superior have put an end to this industry. In western Ontario there are immense deposits of magnetite and hematite, but the lack of railways and of cheap fuel has prevented their exploitation up to the present. A few thousand tons of iron ore have been mined in British Columbia and disposed of in the States to the south, but no iron has been smelted in the province. Vast deposits of ore are reported from Labrador, but at present they are out of reach.

---

*Eng. Mining Jour., 1897.
†Geol. Sur. Can., vol. vii., p. 66 S,

Minor Metals. A considerable number of other metals besides those previously referred to are found more or less extensively in Canada but very little has been done in the way of mining them. Zincblende is found at many points, *e.g.*, near the north shore of Lake Superior. Chrome ore has been mined to the extent of two or three thousand tons during the last two years in the Province of Quebec, the ore being employed in the United States mainly as furnace linings. A beginning has been made in the mining of cinnabar in British Columbia, but no mercury has been produced up to the present. A small quantity of manganese and antimony ores have been obtained in the Maritime Provinces, but they will be referred to at another point.

NON-METALLIC MINERALS.

Coal. Canada is rich in fossil fuels but unfortunately they are not very uniformly distributed, the two most populous provinces, Ontario and Quebec, being almost devoid of them. Coal of carboniferous age is found only in the Maritime Provinces (Nova Scotia and New Brunswick), though barren carboniferous rocks occur quite widely in British Columbia. The other coals of the Dominion are of cretaceous or later age.

There are in Nova Scotia and New Brunswick about 18,000 square miles of carboniferous rocks, but only a comparatively small part of this area contains workable coal seams. Where these occur, however, they are often remarkably numerous and thick, as in the Island of Cape Breton. The carboniferous rocks of the South Joggins on the Bay of Fundy form a classic region for geologists because of the magnificent section displayed, the many trunks of sigillariae and the interesting amphibians sometimes contained in them. The coal seams of New Brunswick seem small and of rather poor quality, while those of the neighbouring province furnish excellent bituminous coal, providing a good quality of coke.

Nova Scotia is the largest producer of coal among the Canadian provinces, its output in 1896 having been 2,235,472 long tons as estimated in the latest report by Dr. Gilpin, in the "Mines of Nova Scotia." Of this 666,403 tons were consumed in the province, 174,919 tons shipped to the United States, and almost all the rest sold in the Province of Quebec, New Brunswick, Prince Edward's Island and Newfoundland. The coal is well adapted for coking and 58,741 long tons of coke were made in 1896.

New Brunswick produces only a small quantity of coal, the statistics for 1894 showing an output of 6,469 tons, most of the fuel used in the province being obtained in Nova Scotia. The interesting coal-like mineral albertite occurs in the province and was mined twenty years ago to the extent of some 200,000 tons. This material is evidently an altered bitumen and has been deposited in a fluid or plastic condition in veins.* The Albert mine provided a most valuable product for the manufacture of gas, but the vein has long been worked out and no similar deposits of economic importance have been discovered.

The Province of Quebec appears to be devoid of coal, its only mineral fuel being peat, which occurs in inexhaustible quantities but has never come into extensive use.

Ontario is almost as poor in solid mineral fuels as Quebec, though small quantities of woody brown coal occur to the north in drift deposits along the Missinaibi and other tributaries of Moose River. A singular anthracitic material has been found in large veins near Sudbury. It has evidently reached its present position, filling fissures in slate of Cambrian age, much as the Albertite of New Brunswick has done; but it has lost

* Geol. Sur. Can., 1876 1877, p. 385, etc.: also 1888 1889, p. 16 T.

so much more of its volatile constituents as to be nearly pure carbon. As it differs in origin from coal, the name anthraxolite, given by Professor Chapman of Toronto University, to small quantities of a similar mineral occurring in Cambro-Silurian rocks, is preferable to the usual name of anthracite or "hard coal." How large a quantity of this fuel is available has not yet been determined.* The important stores of petroleum and natural gas found in southern Ontario will be described in another paper.

The western territories of Canada are richly supplied with coal of late cretaceous or Laramie age, the deposits beginning in the western part of Manitoba and reaching at some points into the Rocky Mountains, the whole area being estimated at about 60,000 square miles. Toward the east this coal is a lignite containing a considerable percentage of moisture so that it crumbles when exposed to the air. As one advances west the amount of moisture diminishes and the lignite approaches bituminous coal in composition when the foothills of the Rockies are reached. Small areas of cretaceous coal nipped in during the elevation of the mountains have the character of semi-anthracite or anthracite proper. Analyses, etc., may be found in reports of the Geological Survey.†

Coal has been mined at a number of points, such as Estevan near the Manitoba boundary, Lethbridge in the foothills and Canmore in Bow Pass, but the only large producer is the Galt mine at Lethbridge. The annual product averages about 200,000 tons.‡ The small quantity of anthracite or semi-anthracite mined in Bow Pass represents the only available supply of that valuable coal in the Dominion.

British Columbia possesses a number of coal deposits all of cretaceous or later age, but up to the present the

*Sixth Rep. Bur. Mines, Ont.
†Geol. Sur. Can., 1882-83-84, M.   ‡Ibid., 1894, p. 19 S.

Nanaimo region on Vancouver Island is the only important producer. The coal from Nanaimo is an excellent bituminous ore, reputed the best on the Pacific coast of North America, and is largely shipped to San Francisco and other points in California. It is said to furnish a good quality of coke. Of the other coal deposits, that of the Crow's Nest Pass in the Rocky Mountains seems the most promising. The area is not large but the seams are numerous and thick, and some of them make excellent coke. It is expected that before long a railway from Lethbridge through the pass will bring these fuel supplies within reach of the smelters of the Kootenay region. In 1894 the coal output of British Columbia was 1,134,507* short tons, of which more than 700,000 went to California.+ In 1895 the output fell to 939,654 tons and in 1896 to 846,235, showing a serious falling off.‡

Asbestus. The Thetford region in the Province of Quebec produces practically all the asbestus marketed in the world. The mineral, though generally called asbestus, is not the true horn-blende asbestus, but really fibrous serpentine, containing nearly fourteen per cent. of combined water. The mineral occurs as small veins in massive serpentine and the silky fibrous parts have to be carefully separated from the rock matter. The material has a number of important uses as non-conducting packing, fireproof material, etc. In 1895 the product was 8,275 tons (of 2,000 lbs.) valued at $347,550.§ The following year it rose to 12,250 tons having a value of $429,856.‖ The total product up to the present is estimated at 84,517 tons worth $5,672,-493.

---

*Geol. Sur. Can., 1894, p. 19 S.
+Rep. Minister of Mines. B. C. 1896, p. 498.
‡Geol. Sur. Rep. in Eng. Min Journal, Mar. 6, 1897, p. 232.
§Mineral Industry, 1895, p. 34.
‖Min. Eng. Jour., Mar. 6, 1897, p. 232.

Mica. A considerable amount of mica is mined, or rather quarried, in Canada, chiefly the amber variety (phlogopite), but partly the white variety (muscovite). Quebec and Ontario are the chief producers of this mineral, which is prepared for stove windows and for the insulation of electrical machinery. For these purposes only those samples which will cut into fair sized sheets are valuable; but of late an excellent non-conducting packing for steampipes is being manufactured from the scrap mica. The product of mica during the past ten years has run in value from about $30,000, to $104,000, the last year's production being estimated at $60,000.*

Minor Economic Minerals. The production of gypsum, apatite, etc., will be referred to by Dr. Ellis under the head of chemical industries, and, therefore, need not be discussed here. Of other minerals few are obtained on a commercial scale. A few thousand dollars' worth of graphite is mined in Ontario and Quebec; a little mineral paint, soapstone, whiting and actinolite are produced, and a few gem stones, though none of high value. Within the past year large deposits of corundum have been discovered in Eastern Ontario, and these will probably be of importance in the future, though probably not as gem material. The labradorites of Labrador with their splendid blue shimmer, the peristerites, grossularites, tourmalines, etc., of Quebec, the jaspers and agates of Nova Scotia and Lake Superior, and the perthites and sodalites of Eastern Ontario afford more or less handsome gem material, utilized, however, to only a very small extent, and valued in 1894 at only $1,600 all told.

STRUCTURAL MATERIALS.

Almost all parts of Canada are well supplied with suitable clay or shale for brick and tile making, lime-

*Geol. Sur. Can., 1891, p. 73 S., and Eng. Min. Jour., Mar. 6, 1897, p. 232.

stone and sandstone for building purposes and the clays and limes used in making cements ; while the Archaean region provides excellent granite, and some varieties of serpentine and marble.

Clay Products. Ontario takes the lead in the production of brick, tiles, etc., and the immediate vicinity of Toronto affords an unsurpassed variety and quantity of clays and shales suitable for making brick of various colors and qualities. The Dominion as a whole is estimated by the Geological Survey to have produced in 1896, brick, etc., to the following value :

Bricks	$1,600,000
Pottery	163,905
Sewer Pipe	153,875
Terra Cotta	110,855
Tiles	225,000

Building Stone, etc. Ontario, Quebec and the Eastern Provinces produce most of the building stone quarried in Canada, and the Eastern Townships of Quebec most of the slate, the values being estimated as follows :

Building Stone	$1,000,000
Granite	106,709
Slate	53,370

Lime to the value of $650,000 and cements to the value of $201,505 are reported in 1896, as well as sand and gravel valued at $120,000.

Conclusion. The total value of the mineral products of the Dominion is estimated by the Geological Survey at $23,627,305 in 1896, metals amounting to more than $8,000,000, non-metallic minerals to more than $15,000,000. The largest single item is coal, valued at more than $8,000,000, next follows gold with a value of $2,810,206, and silver amounting to $2,147,-589 ; while nickel, copper and petroleum, somewhat surpass a million each in value.*

*Eng. Min. Jour., Mar. 6, 1897, p. 232.

21

The two provinces now advancing at the most rapid
rate in the products of the mine are Ontario and British
Columbia, and both appear to have a very bright
future before them, Ontario because of its immense
area of free milling gold ores, British Columbia because
of its great deposits of smelting ores rich in gold, silver,
copper and lead. There is no reason to suppose that
these two provinces or Nova Scotia will prove less rich
in metals than similar areas in the great country to
the south ; and now that Canadians have taken hold
of mining in earnest a great advance may be looked
for in the mineral industries of the Dominion, the law-
abiding character of its people, the wholesomeness of
its climate, the cheapness of supplies and the ease of
access to most of its mining districts affording great
advantages over many other mining regions now
attracting the world's attention.

CHAPTER V.

The Chemical Industries of Canada.

BY W. HODGSON ELLIS, M.A., PROFESSOR OF APPLIED CHEMISTRY
IN THE SCHOOL OF PRACTICAL SCIENCE, TORONTO.

CHEMICAL industry in Canada is only in its infancy. The raw materials for those fundamental chemical industries upon which rests so much of the fabric of modern civilization—acids and alkalies—are abundant although not always conveniently contiguous.

These raw materials are broadly, fuel, sulphur, salt and limestone. All are abundant but hitherto scarcely any serious attempt has been made to utilize them in this direction.

FUEL.

We have in abundance in various parts of the Dominion wood, peat, coal, petroleum and natural gas. Of these coal and peat are discussed above. Wood we have in immense quantity, and it constitutes a large part of our wealth.

According to the last census (1891) there were forty-six establishments for burning charcoal in the Dominion, distributed through the Provinces of British Columbia, Nova Scotia, Ontario and Quebec. The total value of the product is estimated at $91,874. Of this, three establishments, situated in the county of Essex produce more than two-thirds.

Very few attempts have been made to utilize the immense quantity of waste material, such as saw-dust, produced in the manufacture of lumber. Formerly this was thrown into the streams with disastrous results to the fish. Now that this is prohibited by law, the saw-dust is usually burnt. A beginning has, however, been made in the right direction, and saw-dust has been employed in the manufacture of porous terra cotta and otherwise.

The by-products of the charcoal manufacture wood, spirit and acetic acid, are also collected and utilized in some places.

Wood pulp which is now largely used in the manufacture of moulded articles such as pails, and for paper, is produced in British Columbia, New Brunswick, Nova Scotia, Ontario and Quebec, the total product being valued at $1,057,550. The greatest quantity is produced in the Province of Quebec.

Paper is manufactured in Manitoba, New Brunswick, Ontario and Quebec. The census enumerates thirty-four paper mills whose output is valued at $2,575,447.

The bark of the hemlock which abounds throughout most of the Dominion, contains much tannin, and is largely used by tanners. In New Brunswick and Quebec, extract is made from the bark for use in tanning, the yearly value of which is put at $120,000. There were in 1891, 802 tanneries whose output is valued at $11,422,860.

Potash was formerly an important product of the Canadian forests. In spite of the competition of the Stassfurth deposits and of the many modern sources of potash, it is still made to some extent. In 1891, the value of the potash made in Canada was estimated at $153,441, of which about two-thirds was made in Ontario, and one-third in Quebec.

Matches, which may in a certain sense be looked upon as a product of the forest, and which are certainly closely related to fuel, are manufactured largely. There are twelve factories producing matches to the value of $434,953. Seven of these establishments whose output is estimated at $425,253 are in the Province of Quebec.

The development of the petroleum industry in Canada followed hard upon the discovery of petroleum in the United States. In that country Young's process of distilling paraffin oil from bituminous shale was being rapidly pushed forward, and in Canada an establishment was started near Collingwood for the same purpose. Then in 1859, Drake's success in "striking oil" in Pennsylvania put an end to the whole business. At Oil Springs, in the county of Lambton, there existed deposits of inspissated petroleum locally known as "gum beds." When wells were dug to a depth of from forty to sixty feet a bed of gravel was reached just above the rock. This gravel was saturated with oil which rose in the well and was pumped up. In 1861 they began to drill through the rock, and on January 11th, 1862, James Shaw struck a "flowing well" at a depth of 160 feet.* Before October thirty-five of these flowing wells were in operation The quantity of oil flowing from some of these wells was enormous. Mr. Fairbanks calculated † that the Black & Matheson well flowed 6,400 barrels in twenty four hours.

Owing to want of arrangements for controlling this flow much oil was lost. Dr Winchell calculated that "during the spring and summer of 1862, five millions of barrels of oil floated off on the waters of Black Creek." These wells continued flowing about two years, and then the pressure gradually failed, and the oil had to be pumped from them.

About 1865 borings were made in what is now known as Petrolia, about 7 miles north of Oil Springs, and next year flowing wells were struck there also. Some of the wells in this district flowed as much as from 400 to 800 barrels a day. Here, however, arrangements for control and storage were made, and little oil was wasted.

About the same time wells were sunk in Bothwell, about 30 miles away This field, never very pro-

---

*H. P. Brumell Geol. Survey. of Canada, Vol. IV., 1888-89, p. 79, s.

†Mineral Resources of Ontario, p. 159.

ductive, was abandoned about 1865. Last year work
was resumed there, and I am informed by Mr. Blue,
Director of the Ontario Bureau of Mines, that 305,580
gallons were produced up to the end of the year.

Oil has also been found at Manitoulin Island, Pelee
Island, and in many other parts of Ontario ; at Gaspé,
in Quebec : in New Brunswick and Nova Scotia, but
not hitherto in quantities of commercial importance.
It is believed that in the Athabasca region of the
North-West Territories there is an oil field of great
importance.

The petroleum of Lambton is contained in a granular
dolomite, belonging to the corniferous limestone. The
strata penetrated at Oil Springs are 60 to 80 feet of
soil and clay, 160 to 200 feet of the limstones and
shales of the Hamilton formation, and 130 feet of the
corniferous limestone, oil being reached at 370 feet.

At Petrolia the Hamilton limestones and shales lie
about 100 feet below the surface, and extend for about
300 feet. Oil is reached in the underlying corniferous
about 65 or 70 feet down, at a depth of about 370 feet
below the surface.

In the early days of the oil industry, a well about
four feet in diameter was dug to the rock, and boring
continued through the rock by means of a spring pole
and hand power. Skill came with experience. The
soil is now bored with a clay auger 10 inches in
diameter, and the rock drilled with a steel "bit" 4⅜
inches in diameter, and connected with a sinker bar
and "jars" of iron, in all about 38 feet long, weighing
1,250 lbs., and a series of light ash poles attached to
one another by iron screw joints. The drilling is
done by machinery worked by a steam engine. Canadian drillers have a high reputation for skill, and are
in demand in all parts of the world where oil is found.
When the drilling is completed the well is "torpedoed,"
by exploding about 10 quarts of nitro glycerine at the
bottom. None of the wells now are flowing wells,

all have to be pumped, and the average yield for each well is not much more than half a barrel a day. This could not pay if it were not for the ingenious and economical system used. By means of wooden rods or "jerkers" connected to horizontal wheels, set in motion by a steam engine, from half a dozen to a hundred or more wells are pumped by the same engine.

The oil when pumped is run into large tanks, excavated in the impermeable clay which overlies the rock in the district. The tanks are 30 feet in diameter, and 65 or 70 feet deep, holding from 8,000 to 10,000 barrels of oil.

In the neighbourhood of Petrolia there are a number such tanks, whose capacity jointly amounts to 1,000,000 barrels. The tanks are owned by companies who receive the oil from the producers by means of pipe lines, store it and supply it to refiners as required.[*]

According to the report of the Ontario Bureau of Mines for 1895, there were then in operation 6,787 wells in the Petrolia field, and 3,176 wells in the Oil Springs field, making a total of 9,963 wells. The total production of oil was 33,351,997 imperial gallons, giving a yearly average yield of 3,347 imperial gallons per well. In 1891 the yearly average was 6,154 gallons per well. In those four years the number of wells has been nearly doubled, but the production remains nearly the same. The total production of crude oil for 1896 was about 28,000,000 gallons. The falling off may perhaps be attributed to uncertainty as to the tariff.

The oil refineries are now nearly all situated at Petrolia. A large refinery is in course of construction at Sarnia. The stills used are cylindrical iron stills, set in brickwork, and fired by a spray of petroleum. I am informed that the increased value of the heavier products of distillation is causing this method of firing to be abandoned in favour of coal.

The vapours are condensed in pipes, set in long

*James Kerr, Toronto Mail, 1st December, 1888, quoted by H. P. Brumell, loc. cit.

wooden tanks filled with water. The scarcity of the water supply at Petrolia has been a serious drawback to the refiners. The products of distillation are :

1. Gas and uncondensed vapours of very low boiling point.

2. Naphtha, specific gravity, 0.686.

3. Benzine, specific gravity, 0.729.

4. Illuminating oil, specific gravity, 0.788 to 0.082.

5. Intermediate oil, used for enriching water gas.

6. Paraffin oil, from which solid paraffin is crystalized by cold, and separated by filtration under pressure. From the residue lubricating oils of all kinds and of excellent quality are manufactured.

The illuminating oil is refined by treatment with sulphuric acid, which is removed by washing with water and caustic soda. Besides this washing with acid which is required also by the Pennsylvania oil, the Canadian oil has to undergo further purification in order to remove the sulphur compounds which it contains, and which give to the crude oil an extremely offensive smell. This is usually effected by agitation with lead oxide dissolved in caustic soda. The refineries have now succeeded in removing the sulphur very completely from the oil, but the cost of this additional treatment increases the expense of refining.

Mr. Blue, of the Ontario Bureau of Mines, has kindly given me the statistics of oil production for 1896, as follows :—

Crude oil used for fuel	2,221,349	gallons.
Crude oil distilled	25,159,239	"
	27,380,588	"
Illuminating oil	11,342,880	"
Lubricating oil	2,283,047	"
Other products	7,821,262	"
	21,447,189	"
Paraffin	1,532.671	pounds.

The value of these products is estimated at $1,884,430.

Natural Gas.

There are two areas in Ontario where natural gas is obtained in considerable quantity—the Essex field and the Welland field. Their development dates from about 1890. In the report of the Ontario Bureau of Mines for 1892, it is stated that there were seventy-three gas wells in that district producing gas valued at $160,000. In 1896 there were 141 producing wells, 287¼ miles of piping for the distribution of the gas. The production was 2,208,784,000 cubic feet valued at $276,710.

A very large portion of the gas is used in the cities of Buffalo and Detroit.

Sulphur.—We have no deposits of free sulphur of any consequence, so far as known, but in the form of pyrites we possess large quantities which we have scarcely begun to use.

In 1894 (Geological Survey 1895, sec. 169) the production of pyrites was 40,527 tons valued at $121,581. Most of this was shipped to the United States, but a little was used in Quebec and Ontario for the manufacture of sulphuric acid. In 1889 the Petrolia oil refineries used 3,638,704 pounds of sulphuric acid which was made in Canada from Canadian pyrites (Geological Survey 1888-89, sec. 88).

Salt.—There is a large deposit of salt in the Province of Ontario, along the shore of Lake Huron, from Kincardine to Windsor. In 1896 there were made in this district 44,816 tons of salt valued at $204,910. A well bored last year at Windsor is 1,672 feet deep. It passes through four beds of rock salt, the combined thickness of which is 392 feet. The salt from all this district is of excellent quality.

A small quantity of salt is also made in New Brunswick.

Limestones.—Limestones of the best quality, suitable for all purposes, abound in Canada. The value of the lime burnt in 1891, was: in British Columbia,

$40,260; in Manitoba, $261,695; in Nova Scotia, $71,819; in Ontario, $778,550; in Prince Edward Island, $85,844; in Quebec, $393,045; and in the Territories, $18,760; making a total of 1,184 kilns producing lime to the value of $1,444,453.

Cements, both natural and artificial, are produced. The value of the cement manufactured in 1891 was $251,175.

Gypsum is quarried in New Brunswick, Nova Scotia and Ontario, to the value of $118,568.

Of chemical manufactures the census states that there are in Ontario 82 establishments; in Quebec 23; and in the whole Dominion 135, whose products are valued at $2,008,100. This does not include the soap and candle manufactories of which there are 95, producing goods to the value of $2,151,910.

Phosphates.—Canadian apatite began to come into notice about 1878. It is very rich in phosphoric acid, containing about eighty-nine per cent. phosphate of lime and seven per cent. fluoride of calcium, and contains very little iron (Hoffmann Geological Survey Report, 1878). The production increased from this date, with some fluctuations, until 1890, when it reached 31,753 tons, valued at $361,045. The greater portion of this was exported to England. Since that date the competition of the phosphates of the Southern States and elsewhere has produced a collapse, so that in 1894 only 6,861 tons, valued at $41,166, were produced.

The manufacture of fertilizers from this and other materials for home consumption is, however, carried on to some extent. In 1891 there were fourteen establishments for the manufacture of superphosphates, whose output was valued at $247,469.

CHAPTER VI.

AGRICULTURE.

SECTION 1.

General Account of the Agriculture of Canada.

By WILLIAM SAUNDERS,
DIRECTOR OF THE DOMINION EXPERIMENTAL FARMS.

IN Canada, where about forty-five per cent. of the entire population make their living directly from the products of the soil, agriculture overshadows in magnitude and importance all other interests. In the pursuit of agriculture success depends largely on favourable conditions.

A fertile soil is one of the first needs, and in this respect Canada is greatly favoured, having an almost unlimited area of land well adapted to the growing of grain and other crops. The success of agriculture also depends much on climate. The great water system of lakes and rivers affects favourably the climate of the older provinces, and at the same time affords facilities for the transportation of products. The climates of the western area although varying much in character are in large districts favourable to the raising of stock and to the growing of wheat and other cereals of the highest quality. Altitude is also an important factor influencing climate, and in this respect Canada has advantages when compared with other countries. While Europe is said to have a mean elevation of 671 feet above sea level, and North America 748 feet, the altitude of the Canadian part of North America is placed at 300 feet.

PRINCE EDWARD ISLAND.

Prince Edward Island, the smallest of the Provinces of Canada, has an area of 2,133 square miles, of which about 800 square miles is still in forest and woodland. This island has a moist, cool climate in summer, and although the winter is milder than in many parts of the adjacent provinces the cold is at times quite severe. The total precipitation in rain and snow is from 35 to 40 inches. Agriculture here is the paramount industry, a very large proportion of the population being rural. The soil is loamy and fertile, and of a bright red colour having been produced by the disintegration of a soft red sandstone. Until recently the chief products were oats, wheat and potatoes and smaller proportions of barley and buckwheat. Turnips and hay are also important crops. Horse breeding has been carried on to a considerable extent, but owing to the decline in value of these animals this trade has fallen off considerably. Cattle and sheep are kept in increasing numbers: eggs also are produced in considerable quantities. The large yearly shipments of oats and other grain, and potatoes in the past, have gradually resulted in the impoverishment of the soil, but during the past three years the dairy industry has been wonderfully developed and with the number of cattle largely augmented, and the coarse grains and roots with much of the hay fed on the farms, the increased quantity of manure obtained is bringing about a steady improvement of the soil. Fruit growing is not an important industry, but it might be extended with profit. Apples are grown on the island, but not to any large extent; plums also and small fruits yield well, but as yet have been grown only in small quantities.

NOVA SCOTIA.

On the opposite side of the Northumberland straits lies the Province of Nova Scotia with 20,550 square

miles of territory, of which nearly one-third is covered
by forest and woodland. Chains of lofty hills intersect
different parts of Nova Scotia, and in many instances
the lower levels between these ranges are very fertile
and yield excellent crops of cereals and fruit. Although
liable to considerable changes in temperature the climate on the whole, considering its northern latitude, is
temperate. The annual rainfall is a little more than
in Prince Edward Island. Here also the principal
crops are oats, wheat and potatoes, with smaller proportions of barley, buckwheat and rye. The trade in
cattle, sheep and swine is large, and the yearly output
of cheese and butter has of late been much increased.
Fruit growing has also developed within recent times
to a remarkable extent. The Annapolis and Corn-
wallis valleys are specially adapted by climate and
situation for the growth of large fruits, and the choicest sorts of apples, pears, plums and cherries are produced in abundance. Most of the small fruits also
grow here luxuriantly. There are many other localities in Nova Scotia where fruit growing is carried on
successfully and the exports of fruit from this province
are large and constantly increasing. During the past
season the quantity of apples exported was about
500,000 barrels, the greater part of which was sent to
the large cities in Great Britain.

NEW BRUNSWICK.

New Brunswick, which adjoins Nova Scotia, has an
area of 28,000 square miles, of which about one-half
is in forest and woodland. The surface of the country
is generally undulating, but in the northern and north-
west sections there are many ranges of hills, some of
which rise to a height of from 1,200 to 2,000 feet.
Much of the land is rich and fertile, and, when well
cultivated, yields good crops of grain. The climate is
less temperate than that of Nova Scotia, and more
subject to extremes of heat and cold. On the dyked

and intervale lands large crops of hay are grown. The
varieties of cereals and roots cultivated in this province
are similar to those grown in Nova Scotia, with a larger
proportion of buckwheat. The climate is less favourable for fruit growing, and while orchards have been
successfully established in the valley of the St. John
River, and in some other localities, the varieties grown
are chiefly of the hardier sorts, and the cultivation of
large fruits is not general. Small fruits, however, are
grown in abundance, and the cool weather in early
summer retards the ripening season, and permits of
the growing of large quantities of excellent strawberries, which, ripening after the main supplies obtainable elsewhere have been consumed, find a ready
market in the New England cities, at good prices.

QUEBEC.

The Province of Quebec has an area of 227,000
square miles, of which more than one-half is in forest
and woodland. The surface of the country is very
varied, with ridges of mountains and lofty hills,
diversified with rivers and lakes. The climate of
Quebec varies much in different parts of the province,
but the winter weather is steady, and the atmosphere
clear and bracing, with a good depth of snow which
gives excellent sleighing. The summer is warm and
pleasant, when vegetation developes rapidly. Much
of the country is well adapted for farming, the soil
being loamy and fertile. Hay is one of the principal
crops grown, and this has been largely exported, but
with the recent rapid growth of the dairy industry,
much of the crop is now more profitably fed at home,
and the elements of fertility, taken from the land
during its growth, are restored to the soil with the
manure. The principal cereal crops are oats, wheat,
pease and buckwheat, with smaller proportions of barley,
rye and maize. Potatoes and turnips are largely
grown, and cattle, sheep and swine, kept in increasing

numbers. Tobacco is an important crop in this province, and of a total of about $4\frac{1}{4}$ million pounds produced in Canada nearly 4 million pounds are raised in Quebec. Fruits are grown freely in some of the more favoured districts, and there are many good orchards in the valley of the St. Lawrence. Nowhere does the celebrated Fameuse apple reach so high a degree of perfection as on the island of Montreal, where many varieties of pears also, and plums of fine flavour are grown. In the eastern townships which are noted for the excellence of their dairy products, fruit growing is carried on to a considerable extent, and quantities of apples from this section find their way to Montreal. But on the interior lands, on the north side of the river, only the hardier fruits succeed, and the orchards are few and small.

ONTARIO.

Ontario has a territory of 220,000 square miles, more than 100,000 of which is in forest and woodland. This province has about $12\frac{1}{2}$ million acres of cleared land, of which, in 1896, more than 11 million were in field crops and pasture. Ontario has a wonderful variety of climate, the extremes both of summer heat and winter cold being tempered by large bodies of water. The following are the principal crops, with the quantity produced of each, in Ontario, in 1896 : Oats, 83 million bushels ; wheat, $18\frac{1}{2}$ million ; maize, 24 million ; pease, $17\frac{1}{2}$ million ; barley, $12\frac{1}{2}$ million, with smaller quantities of rye, buckwheat and beans. Of potatoes, $21\frac{1}{2}$ million bushels were grown ; turnips, 70 million ; mangels, 17 million, and carrots, $4\frac{1}{2}$ million. The crop of hay and clover was 2,260,240 tons.

Fruit is grown to a very large extent in this province and the possibilities in that direction, are as yet but imperfectly known. The estimated area in orchard, garden, and vineyard, is 320,122 acres. The number of apple trees of bearing age is about 6 million, while there are 3½ million of younger trees, most of which will soon be in bearing condition.

The yield of apples in 1896, is estimated at 56 million bushels. In the Niagara peninsula and along the shores of the western part of Lake Erie, peaches are grown very successfully, and there are said to be over half a million peach trees planted in those sections of Ontario. Grapes also are grown in large quantities. There are about 3 million of bearing grape vines in this province, producing annually about 15 million pounds of grapes. There are also large orchards of pears, plums, and cherries. Canadian markets are well supplied with home grown fruits, and there is a large and increasing surplus for exportation, which is sent chiefly to Great Britain.

During 1896, more than 2 million barrels of apples were exported from Montreal. Tomatoes also are extensively grown, and a large proportion of the surplus crop is canned and sent to other countries. Among other useful crops grown in this province, are flax, hops, clover and timothy seeds, millet, and tobacco.

Dairying has of late years become one of the most important and profitable branches of agriculture. For the feeding of milch cattle and steers during the winter months, a large quantity of maize is now grown and converted into ensilage; nearly two million tons were raised for this purpose during the past year in Ontario alone. Many horses are produced, although owing to the lower prices obtainable for these animals, there are fewer bred than formerly. The number of milch cows is steadily increasing, and a large trade is done in the production of beef, mutton, pork and poultry. The egg trade also is a growing branch of industry.

MANITOBA.

Adjoining the western extremity of Ontario, is the Province of Manitoba with a territory of 64,000 square miles. This is situated midway between the Atlantic and Pacific coasts. The surface is somewhat level and monotonous, with stretches of prairie intersected here and there by wide valleys in which run small rivers and streams, the banks of the valleys being more or less fringed with trees. In many other localities trees are also found in clumps and belts of varying width. Along the ranges of hills which run across this province from the south-east to the north-west, there are forests of considerable magnitude; the proportion of forest and woodland to the total area, being estimated at nearly 40 per cent. The greater part of the land in Manitoba is covered with a deep rich vegetable mould of great fertility. The principal crop is wheat, of which nearly a million acres were sown in 1896. Oats stand next in area with nearly half a million acres, followed by barley with 128,000 acres. The yield of potatoes last year was nearly two million bushels, there was also a large acreage devoted to flax, rye, and pease.

The climate of Manitoba is warm in summer and very cold in winter, but the air is dry and bracing. Winter usually sets in in the latter part of November, and is nearly over by the end of March, although frosts occur frequently at nights for several weeks later. The quantities of cheese and butter made in this province have of late largely increased. The trade in beef cattle is also increasing in importance; the number of animals exported from Manitoba in 1896, was 13,833; the number of swine raised is also greater than formerly, and mixed farming is becoming more general. The climatic conditions are unfavourable to the growth of the larger fruits; but many of the small fruits are produced in abundance.

THE NORTH-WEST TERRITORIES.

Westward from Manitoba, lie the four provisional territories of the Canadian North-west.

Assiniboia has an area of 88,000 square miles; Saskatchewan, 101,000; Alberta, 105,000: and Athabasca, 103,000. These great divisions extending from the western boundary of Manitoba to the Rocky Mountains, are partly traversed by railways, which have opened up the country for settlement and a sparse population of from 50,000 to 60,000 people is scattered here and there throughout this area. In this immense territory there are about 250 million acres of land, a large part of which could be utilized for farming purposes. Up to the present time, less than seven million acres have been brought to the use of the farmer, and about one million are occupied by the ranchers.

Broad and level or rolling plains characterize the territories along their southern boundaries and a wide belt lying north of the 49th parallel—which forms the boundary line between the United States and Canada —extending from about the 102nd parallel of west longitude to the base of the Rocky Mountains, has a dry climate, caused probably by the hot winds which blow northward from the great American desert. This portion of the territories is estimated to contain nearly fifty million acres. The soil of this more or less arid region is, as a whole, of a very fertile character, consisting mainly of a rich alluvial loam, broken in places by tracts of sand and gravelly ridges. The annual rainfall over this dry region varies at different points from seven to fourteen inches, the mean for the whole area being given as 10.91 inches, the larger part of which falls in May, June and July. With some additional water supplied by artificial means much of this land would produce good crops. In the eastern part of this region the supply of water available for irrigation is very limited, but in the western part there are a number of rivers and creeks which are maintained by the melting snows on the eastern slopes of the Rocky Mountains, which in their course offer many favourable opportunities for the diversion of their waters into irrigation ditches for the watering of considerable areas. A staff of engineers belonging to the Department of the Interior has been employed by the Dominion Government for some years past in carrying out a systematic irrigation survey, and information has thus been furnished to the settlers as to the elevation of the beds of the streams at different points and the quantities of water available for irrigation purposes during the summer months. The settlers and ranchers have promptly taken advantage of the information supplied, and from a recent report of the Chief Inspector of Surveys, Mr. J. S. Dennis, we learn that the number of canals and ditches which have been constructed and are now in operation is 115, the mileage of constructed ditches 230, and the number of acres susceptible of irrigation from these ditches is 79,300. In addition to this the number of ditches and canals under construction is forty-five, covering a length of 173 miles, and capable of irrigating 84,250 acres.

Surveys have been made by the Government during 1895 and 1896 for the following additional canals: Bow River Canal, length forty miles, area capable of irrigation 300,000 acres; St. Mary Canal, length fifty miles, area which may be irrigated 50,000 acres, and Red Deer Canal, forty-seven miles, with a capacity for irrigating 50,000 acres.

With a sufficient water supply very large crops of fodder and grain can be grown in most parts of this dry district. The extension of the irrigated area will offer greatly increased facilities for raising cattle and horses, and will also afford sustenance for a large population.

Beyond the spent force of the hot currents of air, beginning from 125 to 175 miles north of the international boundary, immense partly wooded districts are found, watered by streams flowing northward, where the soil is wonderfully rich and fertile, with conditions favourable for mixed farming, and especially for the raising of cattle and for dairying. There the natural grasses grow far more luxuriantly than on the open prairie southward, while the belts and clumps of wood, interspersed with stretches of open country, afford favourable conditions for the growing of grain, with good shelter for stock. The climate in the territories north of the dry belt is much like that of Manitoba, and spring opens about the same time from the Red River to the Athabasca.

BRITISH COLUMBIA.

The most westerly Province, British Columbia, includes 382,000 square miles of territory. West of the coast range the climate is mild and genial, much like that of many parts of England, where the holly, laurel, rhododendron and the yew, flourish with the apple, pear, plum, cherry, and in some districts the peach. In those parts of the province between the coast range and the Rockies there are many fine valleys, more or less utilized for farming and ranching. In some of these the rainfall is not sufficient to admit of the successful cultivation of crops without irrigation; there are, however, many mountain streams available for this purpose, and on some of the ranches very fine crops of grain are grown and excellent fruits, especially apples.

GENERAL CONSIDERATIONS.

The progress made in agriculture in Canada is manifest in many directions. From 1881 to 1891 (the date of the last census) the increase in land under crops was 4,792,542 acres, or more than 30 per cent. The total quantity of land improved and in use by farmers in 1891, was 28,537,242 acres, of which nearly 20,000,000 acres was under cultivation with grain. With the increase in the area under crop is associated a more intelligent system of farming, and greater efforts are made to maintain the fertility of the land. Formerly

Canada was a large exporter of coarse grains, now much the larger part of these crops are fed to animals on the farms. Of about 100 million bushels of oats grown in the Dominion in 1896, there was exported about 1¼ million bushels and about 2 million bushels more in the form of oatmeal. It is much the same with barley. Formerly a large export trade was carried on with this grain, now of the 18 to 20 million bushels. grown not more than 1 million bushels are sent abroad. Thus more than nine-tenths of the whole crop of these coarse grains produced in Canada is now used in the country as food for animals.

The value of the total exports of breadstuffs including all sorts of grain, was in 1896 about 13 million dollars, of which wheat and flour constituted nearly 9 millions. Five years ago these exports were over 27 millions. In the meantime the decrease in that direction has been much more than compensated for by increases in other lines. The exports of animals have risen to over 11½ million dollars, and the exports of provisions which formerly was a comparatively small item now amount to 21¾ million, of which 15½ million are obtained from cheese and butter.

While Canadian farmers are producing annually increasing quantities of grain, these, with the exception of wheat, are being mainly converted into animals and their products, which constitute the principal items of export. The increase in the manufacture of dairy products has been very helpful to the cattle trade, and at the same time it has promoted a rapid development of the swine industry. Pork factories have been established in many parts of the Dominion, and more attention is now paid by farmers to the breeding of those classes of pigs best suited for the production of the highest quality of bacon. By exporting animals and their products in place of coarse grains, the elements of fertility taken from the soil by these crops are largely returned in the manure of the animals and thus the fertility of the land is kept up.

The quantity of land under cultivation in this country, although large in comparison with the number of inhabitants, is very small when compared with the vast areas of rich and fertile country still unoccupied, and in process of time Canada will undoubtedly become one of the chief food producing countries of the world.

CHAPTER VI.—SECTION 2.

Experimental Farms.

By William Saunders,
Director, Dominion Experimental Farms.

THE progress of agriculture throughout Canada has been greatly stimulated by the organization and maintenance of Experimental Farms by the Dominion Government, also by the establishment of agricultural colleges and schools in the provinces, where farmers' sons receive a practical training, and where more or less experimental work in agriculture is conducted. This provincial work will be referred to in a separate chapter. The work carried on by the Dominion Government was begun ten years ago and was designed to cover a very comprehensive field. The Experimental Farms Act outlines the several branches of work to be conducted, as follows :—

" To test the merits, hardiness and adaptability of new or untried varieties of cereals and other field crops, of grasses, forage plants, fruits, vegetables, plants and trees, and to disseminate among persons engaged in farming, gardening or fruit growing, upon such conditions as may be prescribed by the Minister, samples of the surplus of such products as are considered to be specially worthy of introduction.

" To test the relative value for all purposes of different breeds of stock and their adaptability to the varying climatic or other conditions which prevail in the several provinces of the Dominion and in the North-West Territories.

" To examine into the economic conditions involved
in the production of butter and cheese.

" To analyze fertilizers, whether natural or artificial,
and to conduct experiments with such fertilizers, in
order to test their comparative value as applied to
crops of different kinds.

" To examine into the composition and digestibility
of foods for domestic animals, to conduct experiments
in the planting of trees for timber and shelter, to
examine into the diseases to which cultivated plants
and trees are subject, also into the ravages of destructive insects, and to ascertain and test the most useful
preventives and remedies to be used in each case.

" To investigate the diseases to which domestic
animals are subject, to ascertain the vitality and
purity of agricultural seeds and to conduct any other
experiments and researches, bearing upon the agricultural industry of Canada, which are approved by the
Minister."

Five Experimental Farms in all have been established. The Central, or principal farm is located at
Ottawa, where, on the boundary line between Ontario
and Quebec, it serves the purposes of these two important provinces ; the four branch farms being in the
more distant provinces of the Dominion. A site was
chosen for one of these at Nappan, in Nova Scotia, near
the dividing line between that province and New
Brunswick, where it ministers to the needs of the
three Maritime Provinces. One was located near
Brandon, in the central part of Manitoba ; a third was
placed at Indian Head, a small town on the Canadian
Pacific railway, in Assiniboia, one of the North-West
Territories ; and the fourth at Agassiz, B.C., in the
coast climate of British Columbia. At each of these
farms many experiments are in progress in all
branches of agriculture, horticulture and arboriculture,
and many problems of great importance to farmers
have already been solved. In selecting the sites for

these farms, they have been so placed as to render efficient help to the farmers in the more thickly settled districts and at the same time to cover the most varied climatic and other conditions which influence agriculture in Canada.

The Central Farm has about 500 acres of land, with suitable buildings for carrying on experimental work, and residences for the chief officers. There are buildings for cattle, horses, sheep, swine and poultry; also a dairy with modern appliances for experimental work; a seed testing and propagating house, with a building attached which affords facilities for the distribution of large quantities of promising varieties of seed grain which is sent out to be tested by farmers in different parts of the country.

The principal officers of the farm are the Director, Agriculturist, Horticulturist, Entomologist and Botanist, and Chemist. There is also a Poultry Manager, a Foreman of Forestry, who also acts as assistant to the Director, a Farm Foreman and an Accountant. A suitable office staff is provided for conducting the large correspondence both in English and French, which is carried on with farmers in all parts of the Dominion, who are encouraged to write to the officers of the farm for information and advice whenever required.

The Director resides on the Central Farm, Ottawa, and supervises the work on all the Experimental Farms, making personal inspection of the branch farms at least once a year, when the progress of all the divisions of the work is inquired into and future courses of experiments planned.

At the Central Farm, the production of new varieties of cereals and other crops, the ornamentation of the grounds and the plantations of timber trees are under the personal charge of the Director and his assistant, the Foreman of Forestry. During the past seven years more than 700 new varieties of cereals have been produced at the Experimental Farms, by cross-fertilizing and hybridizing, most of them at the Central Farm. Assistance in this work has been had from experts specially employed for this purpose, and also from some of the superintendents of the branch farms. These new varieties are carefully watched and those of less promise rejected.

There are, of these, still under test 189 varieties, viz., 87 of wheat, 33 of barley, 13 of oats and 56 of pease. Some of these new sorts have produced heavy crops for several years and seem likely to occupy a prominent place among the best sorts in cultivation. Many new fruits have been similarly produced, especially of hardy varieties likely to be useful in the Canadian North-West.

About 400 species and varieties of trees and shrubs are being tested in the ornamental clumps and groups on different parts of the grounds, including specimens from all parts of the world where similar climatic conditions prevail. These are carefully arranged in groups and the specimens plainly labelled with their common and botanical names. In this connection there are also large collections of flowering plants such as roses, paeonies, irises, lilies, phloxes, cannas, gladioli, and beds and borders where other attractive perennials are mixed with annual plants.

Twenty-one acres of land are occupied by forest belts, which extend the whole length of the north and west boundaries of the farm and contain about 20,000 trees, including all the more valuable woods which can be grown in this country either for timber or fuel. Annual measurements are taken of the growth of specimens of each variety, and useful data are thus being accumulated. Other objects in view in this work are to determine the results of planting the trees at different distances, also to ascertain their relative growth when planted in blocks of single species, as compared with mixed clumps where many different species are associated together. The value of tree belts for shelter is being investigated, and the most suitable trees and shrubs for hedges tested by the planting of trial hedges of fifty feet in length. Seventy-five of these trial hedges have been planted.

Other branches of the work at the Central Farm in special charge of the Director are the permanent test plots for gaining information on the action of fertilizers on important crops, the seed testing houses and conservatory, and the distribution of seed grain among farmers for trial. Experiments were begun in 1888 with different fertilizers and combinations of fertilizers, on permanent plots of one-tenth of an acre each. The first year these experiments were confined to plots of wheat and Indian corn, but in 1889 the work was enlarged so as to include oats, barley, and roots, and the experiments have been repeated every year since. There are 105 plots now devoted to this work and the results obtained are given in the Annual Report each year.

Arrangements are made every year to test for farmers in all parts of the Dominion, doubtful samples of grain and other agricultural seeds, to determine their germinating power. Last year more than 2,000 samples were tested and their vitality ascertained and reported on; and the information given, prevented, in many cases, the sowing of seed with inferior germinating power.

Those varieties of grain grown on the Experimental Farms which prove to be the best and most productive are annually distributed by mail, free, in bags containing three lbs. each, to farmers in all parts of the country who ask for them. These samples of grain when sown and properly cared for usually produce from one to three bushels, and at the end of the second year the crop will generally furnish the grower with a sufficient quantity of seed to sow a considerable acreage. Such distribution is made at all the farms, but the largest quantity is sent out from the Central Farm.

In 1896, 38,379 samples were mailed to about 35,000
applicants, and during the past seven years more than
160,000 of such samples have been distributed for test
to about 120,000 applicants. In many localities the
new varieties which have thus been introduced are finding much favour and are rapidly replacing the less productive sorts formerly grown. The surplus stock of
seed grain grown at all the Experimental Farms, not required for the free distribution of sample bags, is sold
in larger quantities to farmers for seed.

The Agriculturist conducts the feeding experiments
with cattle to show the most economical rations for
the production of milk and beef. Different rations
have been experimented with, including various combinations of ensilage, roots, hay and straw, with or
without certain quantities of grain. The results have
shown the economy of using ensilage of Indian corn
for the winter feeding of cattle. Many experiments
have also been carried on in the fattening of swine
and much information gained as to the relative value
of the different sorts of cereals for this purpose and the
best methods of preparing them for feeding, also the
usefulness of skim milk, buckwheat, potatoes and roots,
as food for swine.

The Agriculturist also takes charge of the dairy department and superintends the experiments in butter
making. He also conducts experiments with Indian
corn, roots, hay and other fodder crops, testing the
varieties as to their relative productiveness and the
value of different fodder plants and combinations of
fodder plants, as food for stock.

The Horticulturist carries on the work of testing the
relative value of a large number of varieties of fruits
and vegetables on the Central Farm. He also tries
new varieties of fruits to determine their hardiness
and quality, conducts experiments in the production
of new sorts and their propagation, and tests the best
methods of culture both of fruits and vegetables, to

bring them to the greatest perfection. The usefulness
of the various fertilizers in growing these products, is
also tried.

The Horticulturist also conducts experiments in the
treatment and prevention of fungous pests, not only at
the Central Farm, but with the co-operation of fruit
growers these tests are carried on in other parts of
Ontario and Quebec. The effects of varying soil and
climate on fruit and fruit trees within these provinces
are carefully studied and the experience gained from
year to year is published in the Annual Report of the
Experimental Farms or in special bulletins. Specimens
of new fruits are examined and reported on, and meetings of fruit growers in different parts of the country,
attended, where the horticultural work in progress at
the Central Farm is discussed and the experience
gained given for the benefit of those interested. A
large correspondence is also carried on with those
engaged in growing fruit and vegetables. The collections of hardy fruits now in the orchards and small
fruit plantations at the Central Farm are large and
instructive, and each year adds to their number and
interest.

The Entomologist and Botanist to the Dominion Experimental Farms carries on investigations in the life-
histories of injurious and beneficial insects, on the value
of various native and imported grasses for hay and
pasture, as well as on other fodder plants. Particular
attention has also been paid to the subject of noxious
weeds and their eradication. Experiments are also
conducted by this officer, in connection with bee-keeping.

This department is made use of largely by those interested in the scientific aspect of entomology and
botany, many collections of plants and insects being
sent in every year for identification.

In addition to the Annual Reports, where particulars are given of the work done during the preceding

year, timely bulletins are occasionally issued upon entomological and botanical subjects. Successful efforts have been made to get into touch with the best practical farmers and fruit growers in all parts of the Dominion, so as to be informed promptly whenever any outbreak of an agricultural enemy occurs, in order that the best remedies may be applied without delay. By prompt attention to the many correspondents who write to the Entomologist and Botanist, and by the publication of information in the agricultural and daily press, the importance of this department has been made widely known among the farmers of Canada as a source of trustworthy information upon all subjects which come within its scope.

The Chemist analyzes various fodder crops to ascertain their feeding value at different stages of their growth, including native and introduced grasses, Indian corn, etc., and thus gathers information as to the proper time for cutting and curing these plants so as to obtain the best results.

The virgin soils, representing large areas in the Dominion, have been under examination for some years past, and reports given on the analytical and physical data obtained, with suggestions as to the most profitable treatment of these soils. It has thus been shown that Canada possesses soils equal in fertility to the most productive in the world, both among the prairie districts in the North-West and the valley lands on the Atlantic and Pacific coasts.

Many of the naturally occurring fertilizers of Canada —peat, mucks, marsh mud, marl, etc.—have been examined during the past seven years, and much information given in reference to the composition of these important deposits. The knowledge of their composition and value enables the farmer in many parts to enrich his fields at small cost. As far as time permits, analyses are made for farmers, of matters pertaining to agriculture, where the results are likely to be of inter-

est and value to a large portion of the community.
Much useful work has also been done by the examination of water supplies on farms, and by calling attention to the danger of drinking water or of using it for
stock, where it is polluted by drainage from the barn-
yard.

A large correspondence is carried on with farmers
who desire to obtain advice and information respecting
the treatment of soils, the composition and application
of fertilizers, the relative value of cattle foods, and
kindred subjects.

The Poultry Manager conducts experiments with a
number of different breeds of fowls, with the view of
finding out the best egg-layers and flesh-formers, particularly those which give the best egg yield during
the winter season, when eggs are dear. Crosses of different sorts of fowls are made with the same object.
Much attention is also given to the hatching and rear-
ing of chickens, the treatment and food best calculated
to cause vigorous and rapid growth, and record is kept
of the weight and development of the young fowls, per
month, so as to show which thoroughbreds or crosses
give the most satisfactory results and produce birds fit
for market in the shortest time. Notes are also made
as to the behaviour of the different breeds during their
long confinement in winter, and as to the best methods
of preventing egg-eating and feather-picking, also as
to the diseases to which poultry are subject and the
remedies therefor.

When the Central Experimental farm was acquired
sixty-five acres of land were set apart for an Arboretum
and Botanic Garden. This has been placed in charge
of the Foreman of Forestry. During the past seven
years the planting of this section of the farm with
trees, shrubs and perennial plants, has made much
progress, special attention having been given to the
obtaining of as many of the trees and shrubs native to
Canada, as possible, and such species and varieties from

other countries as were likely to prove hardy here. A
large proportion of the native trees have now been
secured, and many of the shrubs and perennial plants;
many species and varieties have also been introduced
from other countries, such as the United States, Great
Britain, Russia, Germany, France, and other parts of
Europe, also from Siberia, Japan, China, the mountain
districts of India, and from Asia Minor. Of these,
many have proved hardy, and the collection already
formed is a source of much interest to botanists, as well
as to the general public. The number of species and
varieties of trees and shrubs now growing in the Arboretum and Botanic Garden is about 2,000, and of
perennial plants, nearly 1,100. These have been arranged, as far as was practicable, in related botanical
groups, so as to admit of convenient comparison and
study. Each tree, shrub and plant, is labelled with a
durable enamelled or a zinc label so that it can be
readily identified.

The Farm Foreman directs the labour of the workmen
and teams, and keeps the time of the men. He also
carries out the arrangements made in connection with
the preparation of the land and the sowing and har-
vesting of the crops, and takes careful records in connection with the growth and yield of all the field crops
and arranges for the storing or threshing, cleaning, and
subsequent care of all these farm products. During
the winter months he arranges for the hauling and
care of manure, the cutting and preparation of food for
stock, and directs the cleaning, hand picking and putting up of the samples of grain sent out to farmers for
test, also of much of the seed supplied to the branch
Experimental Farms.

BRANCH EXPERIMENTAL FARMS.

At the branch farms much of the work is so arranged
as to provide for the investigation of those questions
which are of the most immediate importance to the

farmers residing in the several provinces where these
institutions are located. Each farm is furnished with
suitable buildings and supplied with some of the best
breeds of dairy cattle, also with some of those suited
for the production of beef. Useful tests are made in
all branches of farm and horticultural work, also with
regard to the most practical methods of maintaining
the fertility of the soil.

EXPERIMENTAL FARM FOR THE MARITIME PROVINCES.

At the branch farm at Nappan, Nova Scotia, which
comprises 310 acres, a large number of instructive tests
have been made during the past nine years, particularly
in the growing of oats and barley, and the large
crops obtained of the best yielding sorts have awakened much interest in this subject among the farmers of
the Maritime Provinces. Much attention has also been
given to the growing of fodder plants, grasses, roots
and potatoes, for which the climate is favourable, and
the information gained from the tests made here has
proved valuable. Useful experiments are in progress
in the draining of land, both uplands and marsh, and
the results in crops are showing a marked advantage
as the outcome of this treatment. Experiments have
also been conducted in the feeding of cattle for the
production of milk and beef, and in the fattening of
swine. Orchards have been established and plantations made of ornamental trees and shrubs. The
fruits under test here now number nearly 300 varieties
and the ornamental trees and shrubs include about 280
species and varieties.

EXPERIMENTAL FARM FOR MANITOBA.

The Experimental Farm at Brandon, Man., contains
670 acres ; a part of this land lies in the valley of the
Assiniboine River and a part is on the higher land of
the bluffs. Here much attention has been given to
methods of treatment of land for crops, which has

23

demonstrated the advantage, in that climate, of the
summer fallowing of land, over spring or autumn
ploughing. Different methods of sowing grain have
also been tested and the advantage of the seed drill
over the broadcast machine shown. Grain has also
been sown at different depths, to determine the best
practice in that respect. Experiments have been conducted for the prevention of smut in wheat, a disease
which has been very prevalent in many parts of the
North-West country, and which depreciates the value
of the grain wherever it occurs. The results of these
tests have shown that when the seed is properly treated
smut may be almost entirely prevented.

Experiments have been carried on in the cultivation
of flax, also in the growing of Indian corn, roots, millets
and other fodder crops. In view of the large increase
in stock in Manitoba, and the scarcity of native hay in
some localities, crops of mixed grain have been grown
and cured green for hay with much success. Instructive tests have also been made in cultivating native
grasses and their usefulness proved in the production
of hay. Bulls are kept at this farm, of the most ser-
viceable breeds, for the improvement of stock in the
neighbourhood. Tests have been made in the feeding
of milch cows and steers, to ascertain the most economical methods of producing milk and beef from the
fodder materials most generally available in that
province. Since this farm was established a large
number of the hardiest varieties of fruits have been tested there, and while small fruits succeed well, very little
success has yet attended the efforts to grow the larger
fruits; further experiments, however, are being carried
on in this direction. A large measure of success has
attended the planting of forest trees for shelter and of
ornamental trees and shrubs. Of these, nearly 200
species and varieties have proved hardy and additions
are made to the list every year, showing that there is
an abundance of available material sufficiently hardy

to make successful plantations for the ornamentation of homes in towns and cities, as well as those on
the prairie farms in Manitoba.

This farm, which has been located at Indian Head,
in Eastern Assiniboia, contains 680 acres, and at the
time of its selection it was all bare prairie. The soil is
very fertile and produces excellent crops of grain; but
there is great need of shelter from prevailing winds.
Tree planting on a fairly large scale was begun as soon
as practicable after the farm was occupied, and although at first, progress was rather slow, the trees
planted at the outset soon formed more or less shelter
for each other as well as for those subsequently planted, and now they are nearly all doing well. In shelter
belts, blocks, avenues and hedges, there are now growing on this farm more than 100,000 trees.

Experiments in the treatment of land to prepare it
for crop, in methods of sowing and depth of sowing,
also in the treatment of seed grain for smut, have been
carried on here, the results confirming the conclusions
which have been reached at Brandon. Many tests
have also been made with fodder crops, such as Indian
corn, grasses and mixed grain crops and spring rye cut
green for hay. Experiments have also been conducted
in the feeding of stock, the fattening of swine and the
management of poultry. In this relatively drier climate
the value of good grass for hay and pasture can scarcely be over-estimated, and among the most important
of all the results gained by tests on this farm are
those which have established the value and general
usefulness of Awnless Brome Grass (*Bromus inermis*) in
the North-West. This grass is very hardy, is a strong
grower, endures drought well, makes a very early
growth in the spring and yields fine crops of excellent
hay which is much relished by cattle. Large quantities

of seed of this useful grass have been saved at Indian Head, and many hundreds of sample bags have been sent to farmers in different parts of the North-West country for trial, and the reports received regarding its usefulness are most satisfactory.

Small fruits have been grown successfully at Indian Head ; but among the larger fruits tried none have yet proved successful. Many species and varieties of economic and ornamental trees and shrubs have been tested here and more than 100 sorts have proved hardy.

EXPERIMENTAL FARM FOR BRITISH COLUMBIA.

The branch farm at Agassiz, B.C., is situated in the coast climate of British Columbia, 70 miles east of Vancouver, and contains about 1,100 acres of land, 300 of which is valley land and 800 acres mountain. The climate here is admirably adapted to fruit culture, and since the fruit industry promises to become one of great importance to this province, large trial orchards have been planted on this farm for the purpose of testing side by side the fruits of similar climates, from all parts of the world, so that information as to the most promising and useful sorts may be available to guide the settlers in that country. Already more than 2,200 varieties are under test and orchards have been established not only on the valley lands, but also on the bench lands up the sides of the mountain, at different heights varying from 150 to 1,050 feet.

On the mountain sides have also been planted large numbers of forest trees, especially those representing the more valuable hard woods of the East. Many different sorts of ornamental trees and shrubs are also under test. As at the other branch farms, useful lines of work are carried on in the cultivation of different sorts of cereals, roots and fodder plants, and experiments conducted with cattle, swine and poultry. At this and each of the other farms, many varieties of vegetables and flowers are tested every year, and thus the work is made helpful to agriculture, horticulture and arboriculture.

GENERAL WORK.

During the past seven years more than 10,000 packages of seedling trees shrubs and cuttings have been distributed free through the mail, and more than six tons of tree seeds, in a similar manner, to farmers in different parts of the Dominion who have applied for them and thus a general interest in tree growing has been awakened. An Annual Report is published, containing particulars of the work done at each farm, and this report is sent to every farmer in the Dominion who asks for it. A very large number is distributed each year. Occasional bulletins on special subjects of immediate importance are also issued from time to time, which reach a large proportion of the most intelligent farmers in the country. The officers of all the farms attend most of the more important gatherings of farmers in different parts of Canada, where opportunities are afforded for giving further explanations regarding the work conducted and the results achieved from year to year.

CHAPTER VI.—SECTION 3.

The Work done by the Legislatures of the various Provinces in assisting Agriculture.

By C. C. James, M.A., Deputy Minister of Agriculture for Ontario.

THE legislatures of all the provinces have from the first shown an intense interest in all movements tending to develop this great industry. This interest was first shown in most cases through assistance given to agricultural societies. It will doubtless be surprising to many to learn that agricultural societies were formed as early as 1789 in Nova Scotia and in Quebec, and as early as 1793 in Ontario. Furthermore, all these were inaugurated through the influence of the Governors of the three provinces. In the year 1825, the Board of Agriculture of New Brunswick began the importing of pure bred stock which has been carried on so extensively ever since, especially by the Province of Ontario. To these general agricultural societies were subsequently added the live stock associations. Dairying belongs to a later period. Co-operation in this work dates from 1864, and the various dairy associations belong to the past quarter of a century. All the provinces are at present assisting this industry and the development is very rapid. Another movement of recent years is the formation of societies for discussing agricultural questions as distinguished from societies for holding fairs and awarding premiums. These are known as Farmers' Institutes in Ontario and in Manitoba, and as Farmers' Clubs (Cercles Agricoles) in Quebec. Specialization in agriculture has been assisted by the organization of other associations devoted to fruit-growing, poultry-raising, bee-keeping, entomology, etc. One of the most significant and most hopeful signs of development and growth in Canada is the increasing interest in all the departments of agriculture manifested by the legislatures of the various provinces and the Government of the Dominion.

British Columbia. There is a Department of Agriculture at Victoria presided over by a Minister of Agriculture, who is a member of the Government. Beginning with 1891, an annual report has been issued. This contains reports of crops and live stock from all parts of the province. Some statistics of the industry are collected and published. The legislature makes annual grants to associations and societies in the various districts. Dairying, fruit-growing and stock-raising, receive special attention. There is a Provincial Fruit-Growers' Association to which an annual grant of $1,000 is made, also a Board of Horticulture receiving $1,200. The Department employs a Provincial Inspector of fruit pests. The total grants for agricultural purposes amounted to $14,864 in 1896.

North-West Territories. In this extensive district assistance in agriculture comes mainly from the Dominion Government, the principal grant for this purpose during the present year being $20,000, voted for furthering the dairy interests of the Territories by way of bonuses to cheese and butter factories. Agricultural societies are assisted by both the Dominion and the Local Government. The latter also makes grants to encourage the destruction of wolves, coyotes, and gophers, and to exterminate weeds. Cheese factories and creameries also are assisted by grants or loans at the time of starting. In 1892 there was started The Dairymen's Association of the North-West, the headquarters of which are at Regina.

Manitoba. In the Provincial Government of Mani-

toba the interests of agriculture and immigration are
associated under one Minister. At the present time,
this Minister is also leader of the Government. The
Department collects and publishes full statistical returns of the crops, live stock and dairy products of the
province. The annual report also gives a general
summary of the agricultural work of the association,
and special reports of each give the details. There are
fifty agricultural societies for holding annual fairs or
exhibitions, with one large central association at Winnipeg. $14,000 is voted for the former and $3,500 for
the latter. There are Farmers' Institutes in the various
districts for the discussion of questions relating to
farming. The Central Farmers' Institute, with headquarters at Brandon, is made up of delegates from all
the local institutes. There are three other Provincial
Associations interested in poultry, dairying and stock
breeding. The Government has a Superintendent of
Dairying under whose direction a dairy school is oper-
ated. Special efforts are being made to increase the
number of cheese and butter factories. Other matters
under the direction of the Department are the suppression of noxious weeds, and the inspection of live
animals, with a view of checking and preventing disease. A beginning has been made by the Educational
Department in teaching agriculture in the public
schools.

Prince Edward Island. Agricultural work in this
province has hitherto been carried on mainly by the
Dominion Department of Agriculture. The Provincial Legislature makes an annual grant of $3,000 to the
Provincial Exhibition Association, and $1,500 to each
of two other local associations.

Nova Scotia. Agriculture is not represented in the
Government of Nova Scotia by a Minister, but there is
an officer appointed by the Government to supervise
the agricultural work of the province and to report
upon the same. This officer is known as Secretary of

Agriculture. Nova Scotia possesses one of the oldest agricultural societies in America, the Kings County Agricultural Society in "the land of Evangeline," will in 1897, hold its 108th annual meeting. Agricultural societies have for over a century played an important part in the progress of the province. At present there are eighty-five, with a membership of 4,888, receiving in all $8,000 from the legislature. At Truro is located the Provincial Farm and the Provincial School of Agriculture. At the latter, eighty-seven students were registered in 1896. Dairying is a special feature of the school and special attention is given to courses for teachers. The Nova Scotia School of Horticulture is located at Wolfville, and is directed by the Provincial Fruit-Growers' Association. The school is open from November 1st to May 1st. There is a general provincial organization called the " Nova Scotia Farmers' Association." A report is issued annually by the Secretary of Agriculture giving a summary of the work of all the societies also papers and articles of interest to the farmers of Nova Scotia.

New Brunswick. The supervision of the agricultural work of this province is entrusted to an official of the Government who is known as Secretary of Agriculture. The Provincial Government makes special grants for importing pure-bred stock. Annual grants are made to the local agricultural societies whose funds are used in holding fairs and importing stock for the use of the members. There is a special grant to the fair at St. John, and several local dairy associations are assisted. There is a Superintendent of dairying and a short dairy course is given at Sussex. For the past two years a travelling dairy has been sent out. The leading farmers' organization of the province is known as the " Farmers and Dairymen's Association of New Brunswick," which prints an annual report. The latest yearly report of the Secretary of Agriculture gives the dairy product of the province as follows : In 1895, nine

creameries with 548 patrons, produced 113,892 lbs. butter; 53 cheese factories with 2,292 patrons, produced
1,263,266 lbs. cheese.

Quebec. The work is directed by the Commissioner
of Agriculture and Immigration, who is a member of
the Government, assisted by a Council of Agriculture.
There is an agricultural society in every county. The
Act under which they are organized requires them to
hold exhibitions and competitions for farms or standing
crops in alternate years. The annual grant to these
societies is $50,000. Then there are what are known
as Farmers' Clubs for discussion, one for each parish.
There are 550 of these clubs, and the annual grant to
them is $50,000. The Department sends out lecturers
to attend the meetings of the clubs, each club being
entitled to at least two lectures a year. $6,000 is ap-
propriated for this purpose.

There is issued by the Council of Agriculture the
" Illustrated Journal of Agriculture," which appears
twice a month, once in English and once in French. It
is distributed to members of the societies and clubs,
and its circulation is at present 10,000 in English, and
45,500 in French.

General agricultural instruction is given at five
schools located at Oka, L'Assomption, Compton, Rober-
val, and Ste. Anne de la Pocatiere, the last school
being for housekeeping for farmers' daughters. The
annual grant to these schools is $25,000.

Two veterinary schools are assisted by an annual
grant of $5,000, both located in Montreal, the French
school being attached to Laval University, and the
English school to McGill University.

Two societies for the improvement of Horticulture
are encouraged, viz.: The Pomological and Fruit Growing Society of the Province of Quebec ($500 grant), and
the Horticultural Society of Quebec ($250 grant). In
addition to this $500 is voted to encourage the culture
of fruit trees.

Dairying is especially encouraged in Quebec. The Dairy School at St. Hyacinthe receives an annual grant of $15,000, and $10,000 additional is voted to the Dairy Association of the province and for the inspection of butter and cheese syndicates. Three hundred pupils attended the school in 1896. There were in 1896, 400 creameries and 1,400 cheese factories. The Dairy Association, with headquarters at St. Hyacinthe publishes annually an interesting and valuable report.

Other grants of the legislature that may be mentioned are the following : $200 for the Poultry Association at Montreal, $1,000 for the Official Agricultural Laboratory at Quebec, $5,000 for the Three Rivers exhibition, $4,000 for the improvement of rural roads, and $2,500 for "agricultural merit," this last being awarded in prize competition for farm management.

Ontario. An agricultural society was formed at Newark (Niagara) under the patronage of the first Governor, Col. Simcoe, soon after the organization of the first legislature. It was not, however, until 1830, that organizations of this nature became numerous and received much assistance. The entire Province is now thoroughly organized and districts and township societies are carrying on their work through holding annual fairs and the purchase of pure-bred stock. There are numerous horticultural societies also in the cities, towns and villages. The legislature of Ontario appropriates $75,650 for their support. The sum of $183,736 was paid out for prizes in 1895. There are other provincial societies and associations to which grants are made as follows : Fruit-Growers' Association, $1,800 ; Entomological Society, $1,000 ; two Butter and Cheese Associations, $6,500 ; Horse Breeders' Association, $2,000 ; Sheep Breeders' Association, $1,500 ; Swine Breeders' Association, $1,200 ; Cattle Breeders' Association, $1,500 ; two Poultry Associations, $1,400 ; Bee-Keepers' Association, $1,100 ; and the Agricultural and Experimental Union, $1,200.

Ten fruit experiment stations are maintained in different parts of the province directed by the Fruit-Growers' Association and assisted by a grant of $2,800. Practical instruction in spraying of fruit trees is carried on at an annual expense of $1,800. A vote of $300 is made for experiments in apiculture. The system of Farmers' Institutes is most complete. The Legislature directs the work, maintains the Superintendent, gives a grant of $25 to each separate organization, and sends out well qualified speakers to attend and address the meetings. In 1896, 666 meetings were held, attended by 102,461 persons. The appropriation for this work is $9,900. Road improvement is looked after by an official known as the Provincial Instructor in Road-making. A pioneer farm is established in Western Algoma to prove the adaptability of that section to farming. The Ontario Agricultural College and Experimental Farm is located at Guelph, fifty miles west of Toronto. It is, all departments considered, the best equipped agricultural college in America. Full courses are given leading up to a diploma at the end of a two years' course and a degree (Bachelor of the Science of Agriculture) at the end of a three years' course. Connected with it is a dairy school. Both courses are now attended to their fullest capacity. The Government also maintains a well equipped dairy school at Kingston in Eastern Ontario, and another dairy school with short courses has lately been started at Strathroy. For several years past there have been sent out travelling dairies to give instruction in butter making. The great extent of the dairy production will be appreciated when it is stated that the province produces annually over 100,000,000 lb. of factory cheese (cheddar), 5,000,000 lb. of creamery butter and over 50,000,-000 lb. of dairy butter, and the total value of all the dairy products amounts to over $25,000,000.

All of this varied work is under the charge of the Department of Agriculture at Toronto, over which

presides the Minister of Agriculture, who is a member
of the Government. The reports of all the associations and movements here enumerated are printed and
distributed by the Department. The same Department collects and publishes statistics, including those
relating to the farm and dairy, municipal finances, loan
companies, labor organizations, etc. The appropriation for printing reports and bulletins is $20,500. The
total grant in 1897 for the expenses of the Department,
the grants to the various associations, the maintenance
of the Agricultural College and dairy schools and the
various work coming under the general head of agriculture, amounts to $230,897. The details of the var-
ious branches of this work may be had in the separate
reports published by the Department of Agriculture,
Toronto, Ont.

CHAPTER VI.—SECTION 4.

Notes on Agricultural Education in the Dominion of Canada.

BY JAMES MILLS, M.A., LL.D., PRESIDENT ONTARIO AGRICULTURAL
COLLEGE, GUELPH, ONT.

———

APART from schools and colleges, there are a number
of agencies which assist indirectly in the work of
agricultural education,—agricultural papers and peri-
odicals, agricultural societies, live stock and dairy
associations, horticultural societies, farmers' institutes,
and similar organizations.

In the Province of Ontario, the county and township
agricultural societies, 432 in number, and 51 district
horticultural societies hold annual shows which exhibit
very clearly the results of the best practice of each
locality in grain growing, root cultivation, stock raising, and fruit culture; a provincial fat stock show,
held annually in December, furnishes striking illustrations of what can be done by skill in the breeding,
selection, and feeding of animals; the live stock and
dairy associations have meetings from year to year for
the discussion of questions relating to farm animals
and their products; the latter also send specialists
throughout the province to instruct the makers of
butter and cheese while at work in the factories; and
the Travelling Dairy, under the control of the Agricultural College, goes from neighbourhood to neighbourhood, lecturing and giving practical demonstrations
in milk-testing and butter-making.

The Fruit Growers' Association publishes a monthly
journal and holds annual meetings at different points

in the province for the delivery of addresses and the
reading and discussion of papers on fruits and fruit
culture; the Fruit Experiment Station Board, representing the Agricultural College and the Fruit Growers' Association, conducts experiments in fruit growing
on an extensive scale at ten or twelve different places
in the province, exhibits samples from the different
stations, and publishes an annual report giving the
results of these experiments; and two or three men
sent out by the Minister of Agriculture for a short
time in the spring, go from county to county lecturing
and giving object lessons in spraying fruit trees for
the destruction of insects and fungous diseases.

The Poultry Associations (east and west), the Bee-
keepers' Association, and the Entomological Society
prepare papers and issue annual reports for distribution among the farmers; and the Bureau of Industries
in connection with the Department of Agriculture in
Toronto, collects and publishes from year to year much
valuable information about crops, live stock, wages,
imports, exports, etc., all contributing more or less to
the education of the people in the theory and practice
of agriculture.

FARMERS' INSTITUTES.

Farmers' Institutes are a more important factor than
any of the foregoing in the education of farmers, old
and young, in matters pertaining to their occupation.
The organization consists of deputations of two or
three each sent from place to place to read papers and
deliver addresses on topics relating directly to the work
and life of the farming community—the cultivation of
the soil, the growing of crops, the feeding and management of live stock, poultry raising, bee-keeping, agricultural chemistry, geology, botany, entomology, farm
accounts, practical economics, and many other subjects.

The most important institute meetings in Ontario
are held during the winter vacation of the Guelph

college; and the deputations sent out at that time are
usually composed of members of the college staff and
a few of the most prominent and successful farmers,
stock raisers, fruit growers, etc.   Twenty-one deputations were sent out in 1896, and the number of meetings
held, from one to one and a half days each, was 666.

In this way, every part of the province is visited at
least once a year by the leaders in agricultural thought
and practice—the men who teach the principles of
agriculture and the sciences related thereto, and those
who are most successful in the application of these
principles on the farms of the province.   Thus, much
valuable information is imparted, and farmers are
stirred up to observe, read, and think for themselves.
The work is a great benefit to the country and may
not inappropriately be compared with that done by
university extension lecturers in Europe and America.

ONTARIO AGRICULTURAL COLLEGE.

It is generally admitted, however, that the most
direct and valuable work in the line of agricultural
education in the Province of Ontario is done by the
Ontario Agricultural College, at Guelph.   This institution was founded in 1874, but on account of political opposition and some mistakes in the management,
its progress for the first few years of its history was
slower than might have been expected.   Gradually,
however, it overcame all obstacles, and of late it has
gone ahead very rapidly.   The equipment of the institution at the present time is ample in all departments
—lecture-rooms and laboratories (chemical, physical,
biological, horticultural, and bacteriological); a farm
and dairy supplied with suitable buildings, implements,
and appliances, and well stocked with cattle, sheep,
swine, and poultry; a large garden and a complete set
of greenhouses, with flowers, shrubs, orchards, and
arboretum; and a carpenter shop, with benches and
tools for plain work and general repairs.

The course of study is liberal and very practical, specially adapted to the wants of young men who intend to be farmers. It embraces general agriculture, arboriculture, live stock, dairying, poultry, bee-keeping, chemistry, geology, botany, zoology, entomology, bacteriology, horticulture, veterinary science, English literature and composition, arithmetic, mensuration, drawing, mechanics, electricity, book-keeping, political economy, and German.

The ordinary short course, which is intended as a preparation for life on the farm, extends over two years. Those who complete this course receive diplomas admitting them to the status of associates of the college. Nothing further was attempted for the first thirteen years in the history of the institution; but in 1887 a third year was added for those who should reach a certain standard at the end of the second year and might wish to prepare themselves, not only for life on the farm, but for original work and teaching in agriculture, horticulture, live stock, dairying, and those branches of science which have a more or less direct bearing on agricultural pursuits.

In the early part of 1888 the college was admitted to affiliation with the Provincial University; and since that time all third year work and the final examinations for the degree of Bachelor of the Science of Agriculture (B.S.A.), have been controlled by the Senate of the University.

A large amount of experimental work, of more or less educational value, is now done at the college. A field of fifty acres, divided into about 1,800 small plots, is used for testing varieties of cereals, roots, corn and potatoes, the selection of seed, dates of seeding, kinds of manure, methods of cultivation, etc.; and experiments in horticulture, butter-making, cheese-making, and the feeding of stock, are constantly in progress.

A distinctive feature of the institution is the fact that all students are required to do a certain amount

24

of manual labour while they are getting their education. They are at lectures from 8.30 to 12 a.m.; and for work in the outside departments, they are divided into two divisions, which work alternately in the afternoon, taking their turn at field work, looking after the live stock, and all other kinds of work which may be required in the different departments of the institution. For this work they are paid a certain amount, not exceeding nine cents per hour, which is credited on their bills for board and washing.

The object of this practical work is twofold: first, to assist students in meeting their expenses at the college; secondly, and chiefly, to keep them in touch with the farm and prevent them, during the process of their education, from acquiring a distaste to farm work and farm life—such a distaste as the great majority of students acquire in the high schools and universities of the country.

It may be added that the college has grown steadily in public estimation, till at length it has won the confidence of the farming community. Very large numbers of farmers visit it from year to year (over 19,000 last June), and those who do so generally speak in the warmest praise of the institution and the work done by it. The number of students in attendance last year was 168 in the general course and 69 in the dairy course, or a total of 237.

DAIRY SCHOOLS.

The dairy schools also are doing valuable educational work. There are three of these schools in the Province of Ontario, one in connection with the college at Guelph, and the other two under the control of the president of the college, one at Kingston in the east, and the other at Strathroy in the west. The school at Guelph gives a twelve weeks' course, commencing on the 4th January, and the other two, a succession of shorter courses throughout the fall and winter.

These schools, being maintained by the Provincial Government, are well equipped and well manned, and furnish very thorough courses in the theory and practice of cheese-making, butter-making, milk-testing, the running of cream separators, and the pasteurization of milk and cream. In these courses, farmers' sons and daughters, factorymen and others get, in a short time and at small cost, such instruction and practice as they desire in any branch of dairy husbandry.

OTHER PROVINCES.

While the other provinces of the Dominion have not spent so much money as Ontario directly for agricultural education, or indirectly for the dissemination of information on agricultural topics, they are all working on the same lines.

The Province of Quebec has many agricultural and horticultural societies and six agricultural schools, three private and three maintained in part by public money. The former are at Oka, Wentworth, and Sorel; and the latter, at Ste. Anne Lapocatière, L'Assomption, and Richmond. None of these schools are very strong. That at Richmond, on the Grand Trunk railway, is the best. It has been much crippled by a want of funds, but it has done good work for the people of that locality. The province has also a dairy school at St. Hyacinthe. This is a good school. It gives short courses during the fall and winter and is well attended by both French and English students.

The Province of New Brunswick has agricultural societies, local dairy associations, and a dairy school. The school is situated at Sussex and does good work, but it is open for only a short time each season. The province is divided into three dairy districts, and a superintendent is placed over each. These superintendents give instruction at the cheese and butter factories and hold meetings throughout their respective districts, at most of which they deliver lectures, test milk, and make butter.

The Nova Scotians have a large number of agricultural societies, a school of agriculture at Truro, and a school of horticulture at Wolfville. Both schools are doing well, but they are hampered to some extent by a lack of funds. All public school teachers in training at the Provincial Normal School, which is also at Truro, have to take a course of instruction in the school of agriculture and pass an examination thereon; and a special grant is given to such schools as teach agriculture in addition to the compulsory programme of studies.

Beyond occasional courses of lectures in Prince of Wales College, Charlottetown, Prince Edward Island has not as yet done much in the way of agricultural education.

Manitoba and the North-West are pushing ahead vigorously on the same lines as the older provinces. Manitoba has agricultural societies, farmers' institutes, a travelling dairy, and a dairy school. The school, which is in Winnipeg, is well equipped and is doing good work. Agriculture is taught in the Provincial Normal School and has been placed on the fixed list of studies in the public schools of that young and progressive province—the first and only province in the Dominion which has had the enterprise and courage to say that this important subject must be taught in all its schools.

During the very short time allowed for the preparation of these notes, I have not been able to obtain statistics from British Columbia; but, so far as I can learn, that province has not yet done much for agricultural education.

BULLETINS AND REPORTS.

A large amount of practical information is furnished annually to farmers all over the Dominion through the free distribution of bulletins and reports by the Governments of the several provinces. Last year the Minister of Agriculture for Ontario sent over 40,000 copies each of the bulletins and reports of the college at Guelph and of the principal associations under his control, free of charge, to the post office addresses of farmers throughout the province.

It is not necessary to refer to the educational value of work done by the Dominion Experimental Farms and the Dairy Division of the Department of Agriculture at Ottawa. This, I understand, is fully dealt with in another place.

CHAPTER VII

The Live Stock Industry of Canada.

By DUNCAN McEACHRAN, F.R.C.V.S., V.S., EDIN., D.V.S., McGILL,
DEAN OF THE FACULTY OF COMPARATIVE MEDICINE
AND VETERINARY SCIENCE, McGILL UNI-
VERSITY, CHIEF INSPECTOR OF
STOCK FOR CANADA.

IT may be necessary for the full understanding of
the following remarks by such readers as have
not visited this vast country and acquired thereby
some idea of its greatness in extent and resources, to
preface this article by a few general statements with
reference to the geographical divisions, soil, climate,
and suitability for what may be termed live stock
industry.

By reference to the map of North America it will
be seen that Canada embraces Prince Edward Island,
Nova Scotia, New Brunswick, Quebec, Ontario, Manitoba, the North-West Territories consisting of Assiniboia, Saskatchewan, Alberta, Athabasca, and the vast
unexplored territories lying north of these, extending
as far as the North Pole), and British Columbia. To
travel from the Atlantic to the Pacific Ocean, say,
from Halifax to Victoria, one must spend ten days
and nights on the Canadian Pacific Railway and cover
a distance of 3,756 miles. It necessarily follows that
in a territory so vast a great diversity of soil and climate must exist, and that the breeds and methods
adopted in handling of live stock will vary also.

Prince Edward Island has a mild climate and
moderately rich soil admirably suited for mixed farming. Considerable attention has been paid to stock-

raising, especially horses, both stallions and mares of improved breeds: Thoroughbred and Standard bred have been imported and the island is justly famed for its good horses. Dairy cattle, especially Jerseys and Ayrshires, have also been imported and dairying is now extensively carried on. Agriculture is well advanced generally on the island.

Nova Scotia, in soil, climate, and people, resembles very much old Scotland herself, particularly the western coast of the old land. The soil is excellent in places, and along the rugged sea coast it is specially adapted for sheep grazing. Standard bred horses and thoroughbred crosses are raised. The cattle as a rule are unimproved, although large numbers of Jerseys are to be found in some districts.

New Brunswick, like Nova Scotia, has an extended sea coast but has also much of the finest agricultural lands which are capable of growing unlimited feed for stock. In certain valleys there are marsh hay lands which produce enormous crops of natural hay so cheaply that cattle feeding could not fail to be profitable were these meadows properly utilized for this purpose. The climate and people are essentially Scotch. Considerable attention has in the past been paid to stock-raising, and valuable importations of horses cattle, sheep, and swine have been made by the Government and sold by auction for the improvement of stock in the province. Sheep-raising forms a considerable industry in some districts.

Quebec. This province extends over such a large area that considerable diversity of climate and live stock as well as people are found within its limits, from the barren shores of Labrador to the fertile valley of the Ottawa. The history of stock-raising in this province leads us back to the middle of the seventeenth century during the French régime when the colonization of New France engaged the attention of the Government of Louis XIV., but more particularly certain

zealots of the church, who, to ensure the success of the
colony, found it necessary not only to send the pioneers
and soldiers of France to subdue the native Indians,
but to furnish them with wives which were sent out
in ship loads, and with them, agricultural implements,
seeds, and live stock. Judging from the distinctive
markings of the descendants of the cattle then imported, they must have been brought from the Channel Islands, for in the back country of this province
even to-day the cattle show unmistakable characteristics of the Brittany or Jersey breeds.

The horses imported by army officers would appear
to have been Barbs from the Levant and a few Nor-
mans from Normandy. Probably also at an early date
both cattle and horses were procured from the Maritime
Provinces, where they were imported from the New
England States to the south of them. Even to this
day the native French Canadian horse of this province
shows the head and quarters of the Barb: is of small
size and is endowed with great endurance, probably not
excelled by any of the equine species. Unfortunately
this hardy little animal is becoming scarce. Some years
ago an effort was made by the Provincial Government
to preserve these historical breeds of horses and cattle,
but it has only been partially successful, although both
are in active demand by American farmers.

This province at the present day takes the lead in
dairying, as evidenced by her success in sweeping the
boards at the World's Fair at Chicago in 1893. Valuable herds of Ayrshires, Jerseys, Guernseys, and Hol-
steins, besides the native Canadians, are found throughout the province. Beef cattle have not been neglected,
herds of Shorthorns, Polled Angus, Herefords, and Gal-
loways, bred from importations of the very best blood
in Britain, have made the stock of the province celebrated.

Sheep do well in this province and considerable
trade is done with the States and the other provinces
in rams and ewes of improved blood.

The breeders of the island of Montreal and the eastern townships as well as other portions of the province have done much to improve the breeds of horses, Clydesdale especially, but Thoroughbred, Standard bred, Percherons, and, to a small extent, Hackneys, have been imported.

Stock raising in all its branches can be profitably followed in the climate of Quebec, and proximity to the shipping ports and the United States ensures a ready market.

Ontario is also an extensive province abounding in rich fertile areas admirably adapted by soil and climate to all agricultural pursuits. Serials and roots of all kinds yield large crops throughout the entire province. In certain portions, grapes, peaches, quinces, pears, melons, and tomatoes, form staple crops.

The province is principally devoted to live stock and live stock industries. Herds of the purest blood of every profitable breed of horses, cattle, sheep, and swine, are numerous throughout the province. Ontario breeders are prize winners at all the leading agricultural exhibitions in both Canada and the United States.

Money has not been spared in past years in the importation of the best bred stock procurable in Britain and France, and of such is the foundation of Canadian herds to-day. In western Ontario hog raising and corn growing are extensively engaged in.

Ontario furnishes the largest proportion of live stock and dairy products exported from Canada, both to Britain and the United States, as well as furnishing thousands of stockers annually for feeding on western ranches.

Manitoba will ever be the great wheat growing province but, incidentally, must become a great stock raising province as well.

The progress in this direction is marvellous. Shorthorn cattle are mainly bred, and mixed farming must and will be more extensively followed. This province affords unprecedented opportunities for hog raising; out of the millions of bushels of wheat grown every year, large quantities below the export standard may be bought very cheaply, which, if judiciously fed to pigs, would yield handsome profits.

Stock raising in the Territories, especially Assinaboïa and Alberta, is conducted on a wholesale scale on the ranches. It should be remembered that southern Alberta was the home of the buffalo, the wild cattle of the plains, where they lived summer and winter, moving north or south under certain exigencies of weather or other causes, but always returning here to pass the winter.

In 1881, the writer formed one of an exploring party which visited Alberta, for the purpose of examining this district as to its suitability for cattle ranching—going via Missouri River and driving across the prairie from Fort Benton, Montana, to Morley, north of Calgary. Finding southern Alberta to present physical features of soil, grasses, water supply and climate such as to indicate it to be a veritable cattle paradise, we purchased in Montana 7,500 head of cattle and started the first ranch (the Cochrane ranch) the same year. The soil presents many diversities: from the bare, windblown uplands, on which the stunted pine and creeping juniper grow, to the rich black loam several feet deep forming the grass-covered river bottoms and valleys. The undulating prairie is broken here and there by deep coulees or ravines and long stretches of level upland prairies, and abounds in springs and streams; and in the foothills clumps of trees, mainly the Douglas pine, cottonwood and willows are found. The Bunch grass (*Eriocoma cuspidata*), Blue joint, varieties of pea vines, wild timothy and other grasses, cover the soil in profusion. The altitude varies from 3,000 to 5,000 feet above sea level.

The bright sunshine, the high, dry, exhilarating air,

the abundant vegetation, and the temperate nature of the climate both during summer and winter, makes this district one of the best stock raising countries in the world, where horses and cattle winter out with a small per centage of losses, and these only at intervals of several years.

The explanation of this condition of things will be found in the proximity of the grazing belt to the high ranges of the Rocky Mountains, and in its position on the lee side of these mountains relatively to the prevailing westerly winds. When strong, as such winds often are, they are generally recognized as "Chinook winds," or "Chinooks." In summer they are cool, but in winter carry with them warmth derived from the Pacific Ocean. In ascending the western slopes of the mountains, the greater part of their abundant moisture is precipitated as rain or snow, so that when they flow down again on the eastern slopes there is no moisture present to take up the increase of temperature caused by the condensation of the air. They thus reach the grazing belt in winter as warm, dry winds, absorbing the snow and carrying it off by evaporation without producing (except in rare instances and when accompanied by strong sun heat) any running water, as is usually seen when snow is melted by sun heat at low altitudes. This very condition which renders winter grazing practicable there; renders summer crops very precarious, and, as a matter of fact, the ripening of cereals is the exception, and crop growing in the grazing belt has in consequence been almost entirely abandoned; yet some are hopeful that irrigation may help them to combat nature successfully, which no doubt will be the case if their efforts are restricted to growing cattle food.

Stock raising on the ranches differs materially from farming in settled districts.

The cattle are turned loose on the prairie, with about six bulls to the hundred breeding females, and are handled twice a year only, viz., at the general round-up and calf branding, and at the fall round-up. Hay sufficient to feed the weaklings of the herd is provided to be used only in case of necessity. Calves are branded at the end of June and during September. Beef is shipped during September, October and November. Most of it goes alive to Montreal for exportation to Britain.

A considerable number of cattle go west to British Columbia, especially to the mining districts.

The building of the Crow's Nest Pass Railroad, it is expected, will increase this market.

The men who look after the cattle are called "cow boys;" they are usually men of middle age, of good physique, and capable of much endurance; in manners and ideas of the world in general they resemble very much Jack, the sailor, and like him they are generous to a fault with one another, and have a similar tendency to have an occasional "blow out" during which they usually "blow in" all their month's wages.

There is probably no class of men as much misrepresented as the "cow boy," "cow puncher" or "cow man," of the western country. Judging from the newspaper reports and the manufactured articles of magazine contributors one would suppose that they are all devil-may-care, reckless-of-life, bold-bad-men, whose sole object in life is to ride a bucking broncho, terrorize the natives, "paint the town red," and die with their clothes on.

It is true that the far west has long been the hiding place of fugitives from justice and many lawless outcasts, who dare not return east, work at intervals during the round-up, on cattle ranches, but too often join with others of their own description and become outlaws out west also. These men by adopting the picturesque dress of the genuine cow boy are taken for such and thus despoil him of his good name.

There are to be found on western ranches, wearing the sombrero hat, the buckskin shirt, leather "chaps"

and the clanking spurs, men in whose veins course
Britain's best blood, and graduates of her great Universities, men who, loving a life of freedom, or seeking
to find congenial employment with prospects of financial
success, have adopted the free life in the open air of
the prairie stockman.

British Columbia. The live stock of British Columbia in early days were mostly obtained from the States,
with the exception of a few thoroughbred stallions
imported by the Hudsons Bay Company.

Except a few herds of well bred cattle, the attention
of stockmen in this province has been largely confined
to breeding for beef production. There are several
large ranches of which the Douglas Lake and B. C.
Cattle Co., are the largest.

Hitherto the meat supply of this province outside of
provincial production was obtained from Washington
and Oregon Territories in the United States, now it is
largely obtained from Alberta, and as above remarked
the building of the Crow's Nest Pass Railroad will
increase this trade by affording better and cheaper
transport facilities.

The Cattle Quarantine System of Canada. The
year 1876 marks an important epoch in the history of
stock raising in Canada. It was during that year that
exportation of cattle to European markets may be
said to have actively commenced and the first steps
towards the establishment of cattle quarantines were
taken by the Government. The writer being aware of
the prevalence of contagious pleuro-pneumonia, and of
foot and mouth disease in Britain, urged on the Government through representations made by the late
Hon. George Brown, late Hon. David Christie and
Hon. M. H. Cochrane, all three prominent importers
and breeders of Shorthorn cattle, the necessity of immediate action being taken to prevent the introduction
of these plagues which were decimating British
herds and bringing ruin to British farmers. Thanks to

the clear-sighted practical statesmanship of the late
Hon. Alexander Mackenzie, who was Premier at the
time, an order-in-council was passed under which the
writer was authorized to organize on a small scale
what has since become the most extensive, if not the
most important, live stock quarantine system under
British Government, including as it does every province and territory in the Dominion. Its administration is in the hands of the Minister of Agriculture, a
chief veterinarian and about a hundred inspectors.

Their responsibilities are very great; charged as they
are with preventing the introduction of disease of
animals from without and dealing with all local
diseases of animals occurring within the vast limits of
the Dominion.

Quarantine stations on the sea-board were established
at Charlottetown, Prince Edward Island ; Dartmouth,
opposite Halifax, N S.; St. John, N.B.; and Point
Levis, Quebec, for Atlantic; and at Victoria, B.C., for
Pacific importations. All cattle were till February,
1897, when, in accordance with an agreement entered
into between the Governments of the United States
and Canada, the quarantine of 90 days was removed
and inspection substituted.

The advantages of these quarantines were amply
illustrated in 1886 when contagious pleuro-pneumonia
was discovered in two different herds of imported
cattle and such were the isolation arrangements and
the manner of carrying out the quarantine that the
disease did not extend beyond its walls and Canadian
stock raisers were saved, at trifling cost, from the lung
plague which ruined the stock raisers of Britain, France,
Germany, Holland and Australia, and caused the loss of
millions of dollars to the United States.

Foot and mouth disease was also successfully dealt
with and arrested at the quarantine on several
occasions.

Local diseases such as sheep scab, swine fever,

tuberculosis and actmomycosis, are dealt with by the quarantine staff ; and stock raisers feel assured that no decimation of their herds such as occurs in older countries can take place in Canada.

All diseases occurring in any district of the country are promptly reported to the Department of Agriculture and immediately dealt with under the direction of the chief inspector.

There is, therefore, no stock country in the world, where the health of animals is so remarkable, rinderpest, pleuro-pneumonia and foot and mouth disease do not exist ; the first never did ; the second, only in the quarantine ; and the third, was imported on several occasions both before and since quarantine was established but was confined to the imported cattle, and such are the quarantine regulations that these diseases never can be introduced.

Surely this is a great inducement for stock-raisers of experience to select this country to settle in and prosecute their business, feeling assured they are protected from all risk of animal plagues.

The Exportation of Canadian Livestock. As above stated the exportation of live stock to Britain commenced on a small scale in 1876. In 1877 about 18,-000 head of cattle (mostly bought in Chicago) were shipped from Montreal. The numbers steadily increased till 1893, when the unjustifiable embargo by Great Britain was put on Canadian cattle on the erroneous diagnosis made by Imperial veterinarians, when a non-contagious form of pleuro-pneumonia caused by the rough usage incident to long railroad and ocean transit, was mistaken for the contagious lung plague ; led to a falling off in the numbers exported. In 1896 exportation had again increased to 101,502 cattle, 117,428 sheep and about 15,000 horses in addition to the large numbers exported to the States and sent west to the ranches.

At no time in the history of Canada has the outlook

for live stock business been so promising both in the east and in the west as at present. The removal of the quarantine of ninety days against Canadian cattle by the United States Government after thoroughly satisfying themselves that contagious pleuro-pneumonia does not exist in this country; has opened up again that large market for breeding stock. The opening up of the western mining country, and the shortage, from various causes, of cattle in the far western States tend to boom Canadian ranches, and combine to make Her Majesty's Jubilee year a red letter year for Canadian stock raisers throughout the Dominion.

———

CHAPTER VIII.

Canadian Water Powers.

WITH SPECIAL REFERENCE TO THEIR UTILIZATION
FOR ELECTRICAL PURPOSES.

BY WM. KENNEDY, JR., M. CAN. SOC. C. E., MONTREAL.

THE Water Powers in Canada, suitable for utilization for electrical purposes are so numerous, and of so great capacity, that in a comparatively short article it is possible to do little more than state their locations, or as seems preferable in this case to note some particulars of a few of those now utilized or likely to be in the near future.

In April, 1892, an Act was passed by the Ontario Legislature, confirming an agreement entered into between the Commissioners of the Queen Victoria and Niagara Falls Park, and the Canadian Niagara Power Company, whereby the latter secured from the Park Commissioners the exclusive right for one hundred years, if so desired, to draw water for power purposes at the Falls, and to construct power works within the limits of the Park ; the Power Company to pay to the Park Commissioners $25,000 a year for the first ten years, commencing November 1st, 1892; $26,000 for the eleventh year ; $27,000 for the twelfth year ; and increasing by $1,000 a year up to $35,000 for the twentieth year ; after which the lease is renewable at $35,-000 a year, at the end of each twenty years, up to one hundred years from November, 1892.

By the terms of the agreement the Power Company

was to have commenced construction work by May 1st, 1897, and by November 1st, 1898, they are to "have completed water connections for the development of 25,000 horse power and have actually ready for use, supply and transmission, 10,000 developed horse power." The Power Company endeavoured to have the time for commencing work and the first development of power extended, but this was refused by the Government, whereupon work was nominally commenced, but little has yet been done.

The capital stock of the Company is $3,000,000, with power to issue bonds to $5,000,000.

A water power plant of 2,000 horse power was installed at Niagara Falls five years ago by the Niagara Falls Park and River Railway Company.

Two 45 ins. diameter water wheels, 1,000 horse power each under 55 feet head of water, running 212 revolutions per minute, connect with the generators by means of shafts, gearing and belts. The current so generated is used on the Company's electric railway along the river bank from Queenston, about nine miles below, to Chippewa, about two miles above the Falls. Provision is made for an increase of 1,000 horse power when required.

The Cataract Power Company, of Hamilton, Ontario, has secured from the Dominion Government the right to draw from the Welland Canal, at Allanburg—practically at Lake Erie level—a quantity of water sufficient to furnish continuously 2,000 horse power under 250 feet head. This Company proposes to convey the water from the canal to the Niagara Ridge, a distance of four and a half miles, in an open cut or channel, thence down the slope to the lower level in a seven feet diameter steel pipe to the power house. The discharged water from the water wheels will be delivered into the Twelve Mile Creek, the latter emptying into the old Welland Canal, at St. Catharines, about twenty feet above Lake Ontario level. The water wheels will be directly connected to the 1,000 K. W. generator shafts, running at 400 revolutions per minute. The current will be transmitted to the city of Hamilton, about thirty-two miles distant, at 20,000 volts pressure.

The Chippewa Creek, or Welland River, a comparatively sluggish stream, but with large carrying capacity, empties into the Niagara River about two miles above the Falls. By cutting a channel across country to the brow of the Niagara Ridge, a very large power may be secured from the creek ; the creek bed forming the channel for conveying the water from the Niagara River to the artificial channel, by reversing the natural direction of the flow in the creek, and thus drawing water from the river.

There are several projects in view for the development of large power from the Niagara River, and it is safe to say that when the transmission of power over long distances by electricity, or other means, becomes commercially feasible, all the powers required within, say, one hundred miles from Niagara Falls may be supplied from Niagara River.

All water drawn from Niagara River above, and returned to the river below the Falls, detracts more or less from the grandeur of the Falls, and hence strong opposition has arisen, particularly on the American side of the river, to prevent further concessions being made which will result in diverting any more water from the Falls. On the other hand, the flow over the Falls is so great—estimated at 2,500,000 horse power—that parties who wish to develop power by such diversion claim that the amount of water so diverted, or anything yet contemplated, will produce no visible effect on the Falls.

About one and a half miles west of Rat Portage, and one hundred miles east of Winnipeg, near the westerly limit of the Province of Ontario, the Keewatin Power Company has built a very extensive stone dam

across the Winnipeg river, by which they obtain 20
feet head, and almost the entire flow of water in the
Winnipeg River, with about 3,000 square miles storage
area in the Lake of the Woods. The dam is situated
about half a mile from the main line of the Canadian
Pacific Railway, with provision for running railway
tracks along both sides of the river and connecting
them by crossing on the dam.

When electrical power for general railway purposes
becomes practicable, the Keewatin Power Company
will be in a position to furnish to the Canadian Pacific
Railway Company all the power required ; and with
long distance transmission (say 115 miles) an established success, all the power required at Winnipeg
city, where fuel is very expensive, may be obtained
from the Keewatin Power Company's works,—the
probable capacity of these works being 30,000 horse
power.

On the south side of the St. Lawrence River, near
the city of Montreal, a water power is being constructed
on the Richelieu River, for the Royal Electric Company, of Montreal. The contractors who have undertaken the work guarantee to the Royal Electric
Company a minimum of 20,000 horse power delivered
on the turbine shafts. The head of water is 28 feet.
The power will be divided into eight equal units of
2,000 K. W., and two 500 horse power exciter units.
The water wheels will be placed horizontally in open
flumes, the shafts extending through the bulkhead
walls into the power house and there connected direct
to the generator shafts. These generators will generate
current at 8,000 alternations per minute, and a pressure
direct from the generator terminals of 12,000 volts.
This power will be transmitted to Montreal, sixteen
and a half miles (crossing the St. Lawrence River), and
there distributed for general lighting and power purposes.

The Royal Electric Company claims for this plant,

as distinctive and novel features, the high pressure generated by the dynamos, and that they will be run in parallel.

The Lachine Rapids Hydraulic and Land Company, have nearly completed the first instalment of power at their works at Lachine Rapids, on the St. Lawrence River, a distance of about five miles above Montreal.

The works embrace a wing dam about 4,900 feet long, extending along the line of the Rapids, about 1,000 feet distant from the shore at its lower end and 525 feet at its upper end. About 1,200 feet from the lower end of this wing dam, a dam consisting of 43 masonry piers, forming 36 penstocks and 3 waste weirs, also 4 power house foundations, extends from the wing dam to the shore. In each of the 36 penstocks two vertical turbines will be placed, thus making two rows of thirty-six turbines of two hundred horse power each, under 11 feet head. The turbines will be connected together in sets of six, by means of bevel gearing and horizontal shafting, the latter making 175 revolutions per minute. Each set of six turbines will be directly connected to a three-phase generator of 750 K. W., thus making 12 generators driven by 72 turbines. Power will be transmitted to Montreal at 4.400 volts. There will also be the necessary piers, booms, headracks, etc., for the protection of the works from ice, floating pieces of wood, and rubbish.

The attention of engineers and others has long been directed to the Lachine Rapids as a source of large power convenient to Montreal, the chief commercial city of the Dominion, with a population of 300,000, but owing to the difficulties likely to be experienced by back water, caused by the ice freezing to the bottom of the river and thus raising the surface level of the water at the lower end of the Rapids, and also by floating ice and frazil or anchor ice in the head water, it has not been considered a reliable power, and hence not a safe investment. The promoters and engineers in the present case have, however, not only convinced themselves but have also convinced capitalists that the difficulties mentioned will be overcome by the means which they have adopted. Whether their expectations will be realized or not remains to be proved, but in the meantime their faith has been shown by their very large and costly works.

About 225,000 cubic yards of rock have been excavated, nearly 3,000,000 feet board measure of timber used, over 8,000 cubic yards of concrete, and 6,000 cubic yards of masonry placed.

The capital stock of the company is $2,000,000.

If these works prove to be successful no doubt others of somewhat similar construction will be undertaken, as only about one-fifteenth of the low water flow of the river will be used by the present works.

At Ottawa, the political capital of the Dominion, situated in the Province of Ontario, the great Chaudiere Falls on the Ottawa River, furnishes very extensive water powers. The extreme fall at low water is about 35 feet, and the volume of water about 50,000 cubic feet per second. The quantity is increased about three times during extreme freshets, but the fall is at the same time decreased to about 20 feet by back water from the rise of the river below the Falls. The water has been used for many years for driving very extensive saw mills, pumping the city water supply, manufacturing purposes, electric lighting and electric railways. Probably 6,000 horse power has already been developed for electrical purposes, and no doubt this will be very largely increased as the advantages of power distribution by electricity become more fully appreciated.

At the Deschenes Rapids, on the Ottawa river about six miles above Ottawa City, on the Quebec side of the river, 1,000 horse power has been developed for electric lighting, electric power, and for working an electric railway for regular freight and passenger traffic. This railway, ten miles long, extends from Hull city (opposite Ottawa) to the town of Alymer. It was originally a branch of the Canadian Pacific Railway system, and is the first instance in Canada of an electric railway doing ordinary railway business and exchanging through traffic with steam railways.

At the Chats Falls, on the Ottawa River, 33 miles above Ottawa, there are magnificent water powers in an undeveloped state. The river at this point rushes through channels formed by a number of islands at the Falls. These islands provide excellent sites for power houses. The fall is about 20 feet; distant about four miles from the Ottawa, Arnprior and Parry Sound Railway on the Ontario side of the river, and the same distance from the Pontiac and Pacific Junction Railway, on the Quebec side of the river.

At Montmorency Falls, nine miles from Quebec city, there is an available head of 240 feet of which about 180 feet has been used a number of years by the Montmorency Power Company for lighting the city of Quebec, and power purposes and preparations are being made for furnishing electric power for the Quebec street railway system. The total power of the river at low water is about 3,000 horse power.

The Montreal Cotton Company has recently added 1,000 horse power to their already large water power at Valleyfield, Quebec. The new plant consists of four vertical turbines, divided into pairs, each of which drives a 400 K. W. generator at 200 revolutions per minute by means of shafts and bevel gearing. The main portion of the power used in this mill is furnished by gears, shafting, etc., in the ordinary way.

The water supply is taken from Lake St. Francis, on the St. Lawrence River; the head used is 12 feet, but this may be increased to 15 feet by deepening the tail race. The water power at this site may be increased, say, 3,000 horse power at a reasonable cost.

A water power plant of 800 horse power has been

recently installed for supplying electrical current for power and lighting purposes to Three Rivers, Quebec, nine miles distant. The water is supplied through 375 feet of steel pipe, six and a half feet diameter, to two pairs of horizontal water wheels, 400 horse power per pair, the shafts of which are directly connected to the generators and make 400 revolutions per minute, under 48 feet head.

The power may be increased three-fold when desired.

There are so many large water powers in every province of the Dominion of Canada, extending from the Atlantic Ocean on the east to the Pacific on the west, that the foregoing may quite properly be taken as examples only, rather than an enumeration and description of the water powers in the Dominion, suitable for utilization for electrical purposes, but the space at my disposal forbids a more extended article.

CHAPTER IX.

General Sketch of the Economical Resources, Trade and Population of Canada.

BY JAMES MAVOR, PROFESSOR OF POLITICAL ECONOMY AND CONSTITUTIONAL HISTORY IN THE UNIVERSITY OF TORONTO.

THE first part of this volume and the foregoing chapters of the third part furnish an account of the resources of Canada from the points of view of natural science and of technology. It remains to give a brief account of the economical exploitation of these resources, to indicate the extent of the development of manufactures, transportation and commerce, and the growth and distribution of population.

The foundations of the trade of Canada were laid by the *coureurs de bois*, who trapped and hunted in the "backwoods" of New France, and brought their peltries to Montreal for sale at the periodical fur fairs. The Company of One Hundred Associates (1627), and *La Compagnie des Indes Occidentales* (1664), were the forerunners of "The Governor and Company of Adventurers of England trading into Hudson's Bay" (1670).*

---

* The Charter of the Hudson's Bay Company, granted by Charles II., has been frequently reprinted. (See, *e.g.*, Report of Select Committee on Hudson's Bay Company, 1857, Parliamentary Paper, 224, 260, Sess. 2, 1857.) Much interesting information regarding the history and operations of the Hudson's Bay Company is to be found in "Correspondence, etc., relating to the Northerly and Westerly Boundaries of.....Ontario," Toronto, 1882; in the "Ontario Sessional Papers, Vol. xxi., Part V., 1889;" in "An Investigation of the Unsettled Boundaries of Ontario," by Charles Lindsey, Toronto, 1873; and in "Report on the Boundaries," etc., by David Mills, M.P., Toronto, 1873. "Canada and the States, Recollections," etc, by Sir E. W. Watkin, London [1887], contains an account of the transference of the business of the chartered company to the Hudson's Bay Company, as at present constituted.

In point of area covered by operations the fur trade has always been the greatest of Canadian industries. Even now the trade of the Hudson's Bay Company extends over practically the whole of Canada north of lat. 50°. Their steamers and brigades of boats and carts form the only stated means of communication to the districts of Athabasca, Mackenzie, Keewatin and Ungava (see map at beginning of volume), and the Indian population of these regions is almost entirely employed and supplied by the Company. "Free traders," or "interlopers," buy furs at Edmonton, for example, but few of them penetrate into the northern wilds. Prices of commodities at the Hudson's Bay Company's forts are necessarily fixed arbitrarily and the trade is carried on almost wholly by a system of credits, food and clothing being advanced to the Indians and paid for by the products of the chase.

Much greater in monetary value than the fur trade is the fishing industry. The Gulf of St. Lawrence, the great inland lakes and the Pacific Coast yield immense quantities of fish, both for consumption and for export. Professor Prince's statistics (page 269) give an adequate idea of the relative economic importance and great variety of Canadian fish. By far the most important of these are the cod and the salmon. As regards the cod fishing, it is interesting to note that a system of profit-sharing, coupled, also, with a somewhat dangerous system of credit, prevails on the coast of Gaspé, in the Province of Quebec (as also in Newfoundland). On the Fraser River, B.C., salmon fishing is conducted by 2,300 boats, manned by 4,000 fishermen. The season, which is fixed by law,* extends from July 5th, till August 25th. The bulk of the catch is secured between July 20th

---

* The relations between the Dominion and the provinces as regards fishery legislation has been subject of litigation : see "judgments in Supreme Court of Canada in the Fisheries' Case," Toronto, 1896 ; and *Cf. supra,* Prince, p. 267.

and August 20th. The canning companies lend the
fishermen cash to buy boats, on the understanding
that the fish caught are to be sold to the canneries.
The average price paid to fishermen in 1895, was 20
cents per fish, and in 1896, 25 cents per fish. This
amount is divided among the fishermen in the boat in
prearranged proportions. On the Fraser River, there
are at present 53 canneries. In consequence of the
shortness of the season, and the competition for fish,
the canneries have not only been gradually establishing themselves farther and farther down the river, until
they are now congregated near the mouth, but they
employ steamers which secure the fish from the boats in
the offing. The boats are manned by Europeans, a few
Japanese also being employed, the employment of
Chinese as fishermen being prohibited. In the canneries,
the work is almost wholly done by Chinese labour.
A contract is made with a Tyhee Chinaman for the
whole process of making the fish ready for shipment.
The cans are manufactured and the fish are packed in
cases, the raw material being supplied by the canning
company. Improved machinery by which the cans
may be manufactured is likely to diminish the amount
of Chinese labour required, and also to diminish the
cost of production. The employment of Chinese labour
in the canneries has been the subject of much controversy resulting in demands for the increase of the poll
tax* upon Chinamen. The arguments for the canneries
are that the population on the banks of the Fraser
River does not afford sufficient low-price labour of any
other kind, that ordinary white labour accustomed to
steady employment will not abandon it for a short
season, and that at present it is easier to procure among
the Chinese, expert workmen for the particular service in question—especially for the testing of the cans
when they are filled—than it is to procure such work-

* At present $50.

men among Europeans who go into other, and on the whole more profitable employments. The arguments against the employment of the Chinese are similar to those used in California and in Australia.

In point of area occupied, next to the fur trade and the fisheries, comes perhaps the stock-raising industry, of which an account is given above by Dr. MacEachran. On the great plains of Montana (U.S.A.), Alberta and Assiniboïa, thousands of cattle roam unrestrained over the prairies; the cattle are branded at what is called the "round-up" in the early summer; and in the fall another "round-up" enables the stock-raisers to make their shipments. The economical problems connected with this industry can only be suggested. The relative economy of large and small herds—large herds being from 5,000 to 15,000 head, small herds being from 300 to 400; the relative advantages of keeping a great plain for cattle ranges, and of splitting it up into small enclosures; the relative advantages of settlement with irrigation* and intensive agriculture and ranching; the relative advantages of winter stall-feeding and of the method pursued on the ranges, viz., of allowing the cattle to run as they please during the winter, putting out quantities of hay at certain places, at intervals during severe weather, or while the deep snow prevents the cattle from obtaining the prairie grasses. These, coupled with branding difficulties, and with the difficulty of keeping up the breed, are among the more important questions; the greatest of all being, perhaps, the necessary limit of the endurance of the natural grass under what may readily turn out to be over-stocking of the ranges. At present, however, the shipment of live cattle from the western plains to England by the Canadian Pacific Railway *via* Montreal, appears in an average year to be a fairly profitable enterprise.

---

* On the progress of irrigation in the N.W.T., see the most instructive "Reports of Irrigation Surveys," by J. S. Dennis, Ottawa, 1894, 1895 and 1896.

It has been the practice of the ranchmen to buy young cattle in Manitoba and in the Eastern provinces in order to reinforce their stocks on the ranges. The falling off in the breed of the Manitoba herds owing to the smallness of their numbers, to the inability of the small breeders to improve their stock, and to consequent in-breeding has driven the ranchmen to make their purchases farther east.

The adoption of cold storage and the construction of special warehouses, cars and ships for the conduct of the business, might alter the complexion of the industry, probably to its advantage. The gradual settlement of the country on the main line of the C. P. R. and on its branch lines, has of late years resulted in the driving of the stock-raisers farther a field; and there seems no unreasonable probability that one day the region round Peace River may be as much utilized for ranching as is now the region round the Belly and Old Man Rivers in Alberta.

The comparatively tardy exploitation of the mineral resources of Canada has been well accounted for by Professor Coleman,* who attributes this tardiness to the circumstances of the early settlement of the fertile portions of the country, and to the absence of knowledge of mining on the part of the settler. One result of these circumstances has been that most of the mining which has been done in Canada has either been done by new immigrants from Europe, by miners from the United States, or by Canadians who have learned the business of mining abroad. The mining regions have, therefore, as a rule, a foreign population.† Of late years, however, the attractions of the gold fields, and the pressure upon the professions have

---

Supra, p. 306.

† Witness the case of Lethbridge, Alta. quoted supra, p. 129. The new mining towns in British Columbia—Rossland, Kaslo, Sandon, e.g., are largely populated by miners, prospectors, etc., from Idaho, Montana, and the State of Washington.

caused a great increase in the number both of miners
and of students of mining, with probable important
results in the future so far as the distribution of
population is concerned. Several administrative and
economic problems have emerged in course of the
recent rapid development of Canadian mining. The
maintenance of order has happily so far been an easy
business. In spite of the traditional turbulence of
mining populations and the eagerness with which the
pursuit of gold is conducted in them, the newly developed regions have exhibited remarkable instances of
self-government in the literal sense. It is, however, a
question whether or not this can be a permanent condition. In such a case much depends upon the tact
and sagacity of the government officials on the spot
and upon the good will and practical equality of condition of the miners.*

On the economic side, the principal problems connected with gold and silver mining are those of smelting and of transportation. Many of the most valuable
deposits of placer as well as of free-milling quartz and
of dioritic gold-bearing rocks are at present inaccessible by ordinary means of travel. No doubt, in time
the yield from the less inaccessible mines will justify
the extension of the system of transportation, by
means of tunnels connecting the different mines together, or by means of sub-aerial or aerial railways or
tramways. The remarkable system of inland waters
in the Rainy Lake and Seine River districts, and in
southern British Columbia and in the Yukon have
rendered possible meanwhile a development to which
otherwise Canada might have remained a stranger.

The smelting of Canadian gold, silver and copper
ores, is carried on to some extent in Canada, especially
at Sudbury in Ontario, and at Nelson and Trail, in

---

* An "Uitlander" question has already emerged in the West
Kootenay district of British Columbia, and in the Yukon.

British Columbia. The bulk of the Canadian ore goes, however, to the smelters at Butte, Helena and Great Falls, in Montana, or to Omaha, in Nebraska. Even where the ores are smelted in Canada, the process of refining is carried on in the United States. The bullion exports from Canada are therefore small in comparison to the total yield. The gold and silver extracted from Canadian ores appear as United States products.

Wheat growing in Canada is now chiefly carried on in Manitoba, although there is an increasing quantity grown in the North-West Territories and a diminishing quantity grown in Ontario. The thick, black loam of the prairie west of the Red River has produced, under cultivation, the finest wheat. Although Manitoba wheat commands a slightly higher price in the market than other sorts, the price is determined rather in the " wheat-pit," at Chicago, than either at London or at Winnipeg. The relatively low price of 1895 was offset by an abundant crop; and the relatively high price of 1896 was offset by a deficient crop, due largely to late storms. On the whole, the Manitoba farmers have been prosperous during the few past seasons, and have succeeded, not only in paying interest on mortgages for money borrowed during bad seasons, or to purchase machinery, but have been able to pay off a large amount of their indebtedness. The result has been a check to the inflow of capital into the province.

The trade in wheat is conducted by a few large firms, who have their warehouses or elevators at nearly every station in Manitoba, and at many in the North-West Territories. The farmers take their grain to the elevators and receive the market price, the cost of carriage to the market being taken into consideration.

Although there are large flour mills at Montreal and in Ontario, there is a much larger exportation of wheat than of flour. Among the reasons for this are the expediency of mixing different classes of wheat in flour milling, the absence in Canada, at present, of facilities for the disposal of offal, and the relatively greater convenience of storing large quantities of wheat than of flour. It is to be noted, however, that a market for flour is opening up in Australia. There is an apparent tendency towards consolidation of the flour milling companies.

The Ontario, Nova Scotia and New Brunswick farmer, driven out of the wheat market by the competition of Manitoba, has turned his attention to cattle-raising, to dairying, to fruit and vegetable growing, and to bee culture.* As regards vegetables, there are three staples in which, especially in the Ontario peninsula, a large trade is carried on. These are tomatoes, corn and peas. As regards fruit, strawberries, peaches, grapes, plums and apples are the principal staples. The trade in fresh fruits has been to some extent injured recently by the competition of dried fruits from California. Apples are produced in large quantities also in New Brunswick. A change has recently taken place in the apple trade of interest from the point of view of international commerce. Formerly the whole of the trade in apples between Canada and the continent of Europe was transacted through English and Scotch firms, who transhipped the apples in the United Kingdom. Now the Canadian houses export direct to Germany and France.

The development of the cheese and butter industry has been greatly promoted by the establishment of co-operative creameries and cold storage arrangements; and by the adoption of improved methods by which at

---

* In addition to the information contained in the articles by Mr. Saunders, Mr. James and Mr. Mills above, much valuable information may be had from the reports of the Dominion Minister of Agriculture, and from the reports of the Ministers of Agriculture of the various provinces, especially of Ontario. Information regarding the Dairying Industry, Cold Storage, etc., may be found in the excellent reports of Mr. Robertson, Dairy Commissioner, Ottawa.

a day's notice the management of a factory may change from the manufacture of cheese to that of butter. The result of these recent improvements has been a great increase in the export trade both in cheese and butter.

Pork packing is carried on at Winnipeg and at Toronto, St. Thomas and Ingersoll, in Ontario. Formerly farmers killed hogs on the farm; now they send live hogs to market where they are bought for the packing factories. The conditions of the Canadian trade have lately been such that the farmer found greater profit in the production of light-weight than of heavy-weight hogs, and with the ready adaptability for which the Ontario farmer has distinguished himself, he has devoted himself to the production of these exclusively. For packing purposes, however, it is necessary to have a certain number of heavy-weight hogs, and recently these have had to be imported from the United States.

Boot and shoe factories, so far as Canada is concerned, have come to be concentrated in Montreal and Quebec. There were formerly some factories at Toronto; but the industry has practically left Ontario. The principal cause of this concentration appears to be the cheapness of labour in the lower province, where French-Canadian girls are largely employed.

The manufacture of wood into doors and window-frames, etc., is carried on extensively in Ontario—at Desoronto, for example, the work being done by the aid of wood-working machinery, and large quantities are packed and exported. Furniture, also, is manufactured extensively in Canada. There is a certain freshness of design in Canadian furniture, although there is perhaps a tendency to a rococo style.*

The manufacture of cottons is engaged in more largely in the Province of Quebec than in Ontario.

* Furniture is manufactured chiefly at Preston, Toronto, Ottawa, London and Woodstock.

26

The largest mill is at Hochelaga. The Dominion Cotton Company, one of the most extensive concerns, has ten mills with 194,000 spindles and 4,390 looms. The trade in printed cotton goods has increased enormously during the past five years. In 1890, the deliveries were 87,000 pieces ; in 1896, they were 500,000 pieces.

During the American Civil War, while the cotton famine lasted, and the price of linen was high, a factory for the manufacture of linen was established at Waterloo, Ont., but after the close of the war and the fall of prices, the manufacture was abandoned.

Shipbuilding is engaged in at Owen Sound and at Toronto ; although the decline of wooden shipbuilding has affected this industry at the former port and has affected related industries, such as sail-making, at both places.

Hops are grown in British Columbia—in the Okanagan Valley, for example. Canadian hops are being exported on a considerable scale to Japan, where beer seems to be replacing the native spirits.

There are breweries in Brockville, Walkerville, Perth, Waterloo and Toronto ; and distilleries in Halifax, Waterloo, Lanark and Toronto, and in the counties of Hastings, Grenville and Essex.

Tobacco is grown in Essex and Kent counties, Ontario, and manufactured in Montreal, Toronto and Hamilton.

The Canadian wholesale and retail trade in dry-goods has undergone a series of changes during recent years. The growth of "departmental stores,"* and the efforts made by English wholesale houses to trade direct with the "departmental stores" or with the ordinary dry-goods store instead of selling to the wholesale houses in Canada, have together tended to render the wholesale trade unprofitable so far, at any rate, as the large

---

* Shops, of which Whiteley's, in London, or Lewis's, in Liverpool, may be taken as the type.

towns are concerned.* Even in the country, dry-goods merchants in some cases unite together to send a buyer to New York or London. The efforts toward direct trade on the part of the English houses have led to the appointment of agents or brokers : this is manifest in the grocery trade as well as in the dry-goods trade. The extent of territory which an agent must cover in selling even one commodity and the cost of travelling render it more profitable for him to represent several houses and to sell sometimes widely different classes of goods. The wholesale houses have, however, an advantage in trading with the more distant parts of the country. In spring and autumn, for example, dry-goods buyers from British Columbia and Manitoba go to Montreal, Toronto or New York, for the purpose of seeing the latest *modes* and of laying in a stock of materials with which to make up these.†

The soft coal of Nova Scotia is used for maritime purposes and for manufactures in that province and in Quebec. Western Ontario is wholly supplied from Pennsylvania.‡

The lignite coal beds of Lethbridge supply the Canadian Pacific Railway through a great part of the Territories and the line to Great Falls in Montana. The bituminous coals in the foot hills of the Rocky Mountains have for years been used by the ranchmen in the neighbourhood, and are likely to be

---

* One result of this has been that the wholesale houses in the towns will sell to country retail dealers at a lower rate than to town retail dealers. The latter thus sometimes go to the country villages to buy. The several conditions noted in the text and above (as also relatively high local railway rates), have contributed to the development of country markets, and to a corresponding falling off in the markets at the centres.

† A curious point in connection with buyers' excursions to the central towns in Ontario, is that the wholesale houses in the country towns induced the railway companies to abolish these excursions, on the ground that they injured local trade.

‡ See, however, the possibilities of anthraxolite and other fuels— *supra*, pp. 318, 323.

thoroughly exploited on the opening of the Crow Nest Pass line of the Canadian Pacific Railway, now in course of construction. The anthracitic beds of the Rocky Mountains on the line of the Canadian Pacific Railway supply the coal for that section of the railway. Several hundred thousand tons of soft coal are exported yearly from Nanaimo in Vancouver Island to San Francisco. The labour at Lethbridge is largely foreign.* In Nanaimo, although Japanese are employed at the pit head, none are employed in the mine. Chinamen are not employed.†

The presence of tanning barks, principally hemlock, in large quantities in Canada, has brought about the establishment of tanneries in close proximity to the sources of supply of tanning agents. The chief source of supply of hides is South America. There is a large export trade in sole leather; and this is at present essential to the manufacture, because in making the class of sole leather demanded in Canada (No. 2 medium) it is only possible to utilize about one-half of the hides, which have to be purchased unselected in South America. The heavier class of leather which, under these conditions, has to be produced is exported to England. The conditions of the sole leather tanning trade involve the employment of considerable capital, and facilitate uniformity of methods and terms of credit. The black leather tanners of Canada appear likely to gain under the Dingley Bill in consequence of the duties upon hides and tanning materials imposed by it, in the United States, such materials not being subject to duty in Canada.

<hr>

* See supra, p. 129.

† They were formerly, but owing to a strike of European miners at Nanaimo some years ago, their employment was discontinued. Japanese are employed in the lumber yards at Vancouver. Chinese are not employed. The Chinese population is chiefly employed in trade, gold finding (in the river gravels), in market gardening, in laundries and in domestic service.

On account of the quality of the wool, Ontario has for many years been better suited for the manufacture of worsteds than Quebec. Ontario farmers have established good breeds (largely Shropshires) of wool producing sheep. The wool trade thus obtained a start in Ontario which it has since maintained. In the North-West Territories, as also in Montana, Merinos are bred for the finer wools.* The production of large sheep in the United States has not been sufficient to supply the worsted industry there and wool is largely imported from Canada. On the other hand Canada supplies the deficiency by importation from London of South African and Australian wools.†

The principal centres of engineering industries are Sherbrooke, Que., Hamilton, Owen Sound and Toronto. Mining and other machinery is manufactured largely at Sherbrooke. There are very large electrical engineering works at Peterborough. Dynamos and all the accessories of electric lighting and the utilization of electrical powers are made there in large quantities. Bicycles are made at Toronto and at Brantford. The smaller manufacturers appear to be succumbing to the effective competition of the larger makers, alike as regards selling price and as regards the control of patents.

Among the miscellaneous industries of importance are pressed brick making, in the Don Valley for example, the production of salt at Windsor,‡ India rubber goods at Toronto, piano making at Toronto, and button making at Berlin, Ont.

Sugar is manufactured in Montreal and in Halifax. There are two cane sugar refineries, and one beet refinery. Agricultural implements are made at Toronto. In Manitoba a large quantity of flax is grown for seed

---

* It is a question, however, how far sheep ranching and cattle ranching can be carried on successfully in the same region.

† For many of these details I am indebted to W. H. Moore, B.A., of the University of Toronto.

‡ See *supra*, p. 329.

purposes. The fibre is not suitable for textile manufacture.

The railway system of Canada, partially shewn in the map at the beginning of this volume, extends for 16,091 miles. The number of railways in operation is 138; the number of controlling companies is 74.

There are two Government railways, the Intercolonial and the Prince Edward Island Railway. The Standard gauge in Canada is 4 ft. 8½ in.* Government has expended altogether about $150,000,000 in railway construction, about one-third of this has been paid by means of subsidies to joint stock companies, about one-third invested in railways which have afterwards been handed over to companies, and about one-third in lines which still remain under the control of the Government. Three great lines of railway practically divide the territory of Canada among them. The Canadian Pacific Railway—mileage, including railways in the United States belonging to or worked by the Railway, 7,219; Grand Trunk Railway—mileage in Canada alone, 3,162, in the United States, 1,005; and the Intercolonial (Government) Railway—mileage, 1,136.†

The Grand Trunk Railway (largely built 1852-1857), was at the beginning a commercial enterprise, although it was obviously not without influence upon the political development of the country. The Canadian Pacific Railway and the Intercolonial Railway were both undertaken chiefly from political motives. Not only was their construction insisted upon by the provinces

---

* There is one broad gauge railway, the Carillon and Grenville (5 ft. 6 in.), and several narrow gauge railways (3 ft. 6 in. and 3 ft.). The gauge of the Grand Trunk Railway as originally projected was 5 ft. 6 in. See statements, reports, etc., Grand Trunk Railway, 1857, where also find much information in the early history of the railway.

† Detailed statistics of the Canadian railways are published in the report of the Department of Railways and Canals (Dominion Government), published annually at Ottawa.

concerned; but it was apparent to the statesmen who effected the confederation of the provinces that the bond of a continuous railway system was essential to political union. The Canadian Pacific Railway has, however, rendered the material prosperity of the great heritage to the West, not only possible but certain. It is true that population increases slowly and that immense regions are still uninhabited; but it is not yet twenty years since the wheat fields of Manitoba came into existence, and only fifteen since ranching and settlement even began in the Territories. The construction of the C. P. R. was begun by the Government,[*] and afterwards transferred to the present company, together with a large land grant, and a cash subsidy, with the obligation to construct, equip and conduct the line. The resulting relations between the Canadian Pacific Railway Company and the public are thus unusual. A large part of the line cannot for many years, if ever, be of itself commercially profitable; the line as a whole is a single line of immense length, rather than a system, and the reasons for its length are chiefly political reasons. A large subsidy was therefore inevitable; and the efficiency of the service can only be maintained at considerable cost. As the country becomes settled and the traffic increases, the commercial success of the Company becomes more and more assured.

The interests of the public, of the shareholders of the Railway Company, of the Dominion Government, of the different Provinces, of the different municipalities, of the holders or mortgagees of land at different points, of the settlers, of the traders, rural and central, of the travelling public, are by no means readily har-

---

[*] The construction of the Canadian Pacific Railway is fully described in the reports of the Engineer-in-Chief, Mr. (now Sir) Sandford Fleming. Descriptions of different routes are also to be found in Capt. Palliser's well-known report, and in the reports of the Geological Survey, Ottawa.

monized. Railway rates are, as everywhere, the chief point of attack,* although there is also, from time to time, agitation for the construction of new lines by the Government or by companies with or without Government aid. Hitherto the demand for new lines has been yielded to with moderation, the management of the chief lines realizing that serious risk to the railway companies lurks in too hasty railway building in a sparsely populated country notwithstanding the inducements of free land grants and cash subsidies; while the danger from a national point of view of overwhelming interest charges is very serious.

The inland waterways of Canada are unrivalled in many ways; but they suffer from the disadvantage of great change of level. The numerous waterfalls and rapids make navigation impossible without canals at certain points. Prior to Confederation, the Provincial Governments, with some aid from the Imperial Government, had expended $20,594,000 upon canals. Since Confederation the Dominion Government has expended $44,100,000 out of capital, in addition to $2,225,000 expended out of income; or a total expenditure of nearly seventy millions of dollars (£14,250,000). The policy of the Government has been for some years to secure a minimum depth on the lake, canal and river route from the great lakes to the sea of 14 feet. At present the minimum depth is 9 feet; but in two or three years from now (in 1899 or 1900), it is estimated that the depth of 14 feet will be attained throughout. The new Sault Ste. Marie Canal, which has cost $3,250,000, is one of the finest in the world.†

---

* Discussions of the railway rates question in Canada are to be found in Mr. G. R. Parkin's "The Great Dominion," London (McMillan); in "The Railway Question in Canada," by J. S. Willison, and in the Reports of the Royal Commissions on Railways (1888) and (1895). The Canadian railways enter into joint traffic arrangements with the leading lines in the United States.

† Statistics of Canadian Canals, with descriptions and reports on construction and maintenance, will be found in Annual Reports of the Department of Railways and Canals, Ottawa.

The movements of internal commerce in Canada have been incidentally noticed in connection with particular industries. It is very interesting but impossible, at present, to study in detail, the movements of provincial and inter-provincial trade.* A good harvest in Manitoba results in the despatch principally from Ontario to that province, which at present scarcely manufactures at all, of large quantities of agricultural implements, dress goods, furniture, pianos, wines, etc.; while a poor harvest, or a low price, prevents such importation. A "boom" in gold mining in British Columbia results in the congestion of the railways with passengers, supplies and mining machinery from the east, and in the return traffic of ores to the smelters.

The external commerce of Canada has been subject to considerable fluctuations since Confederation. At intervals of five years since 1868, the following amounts exhibit the total trade (imports and exports, including coin and bullion):—

$000,000's omitted.		$000,000's omitted.	
1868	131	1888	201
1873	218	1893	248
1878	172	1895	224
1883	230		

Dutiable goods in 1868 amounted to $44,000,000, out of a total entered for consumption of $72,000,000. In 1895, dutiable goods amounted to $59,000,000, out of a total entered for consumption of $105,000,000.*

Wholesale prices in Canada cannot be said to have followed an independent course, except in so far as they have been affected by alterations in the tariff. In 1868, the year after Confederation, wheat, barley, and all farm products were high; and the general level

* Report of Department of Trade and Commerce, Sessional Papers, 1896, Vol. XXIX., No. 4, p. 3.

† Interprovincial trade is estimated at upwards of $100,000,000.

of prices was high in relation to the average of the period from 1867 till 1897. The price of wheat fell in the following year to a point below the average, rose in 1871 and 1872, fell steadily till 1875, rose till 1877, fell sharply below the average in 1878, and remained till 1882 at a point about or below the average, falling steadily from 1882 till 1894, with slight recovery in 1888, 1890 and 1891. Since 1894 wheat has sharply recovered. A curve of wholesale prices of the chief commodities would exhibit approximately the foregoing features.

The main lines of the tariff history of Canada * are these :—

1. Prior to the Confederation in 1867, of the colonies which now constitute the Dominion of Canada, each colony had a tariff of its own. The Maritime Provinces adopted the system of specific duties with a low tariff (about 12½ per cent.) for revenue purposes, while the provinces of Canada (Ontario and Quebec) adopted in the main the system of *ad valorem* duties with a relatively high tariff.† One of the difficulties of Confederation was the existence of these divergent tariffs. The settlement arrived at involved the adoption by the confederated provinces of a tariff which approximated rather to the low tariff of Nova Scotia than to the high tariff of the Canadas.

2. The outcome of this situation was that for the first seven years of Confederation (1867-74), the tariff of the new Dominion was about 15 per cent., while the duties were more largely specific than had formerly been the case in the Province of Canada. A period of prosperity followed Confederation, Canada having

---

* An account of the Tariff History of Canada up till 1895 is to be found in the " Tariff History of Canada " by S. J. Maclean, B.A., Toronto, 1895.

† The tariff of Canada of 1858-59 was indeed protective. Against this feature the then Colonial Secretary, the Duke of Newcastle, protested without avail. Maclean, op. cit., p. 9.

shared in the expansion of trade which followed the Franco-German war. In 1871, there was a surplus in the Canadian Treasury of $4,000,000 and in 1872 a surplus of $3,000,000 ;* but in the years succeeding 1875, a series of deficits shewed that Canada had also shared the general reaction.

3. The inevitable result was an increase in the tariff. Still the increase was slight, and was due rather to the necessities of the revenue than to any other motive. From 1874-78 the tariff was about $17\frac{1}{2}$ per cent. Meanwhile, however, the increase of population, the need of large public expenditures in order to implement the engagements into which the Dominion had entered with its constituent provinces upon the confederation of these, together with the rising manufacturing interest especially in Ontario, lent weight to the demand for a still higher tariff. From 1875 onwards this demand assumed a more and more definite form.

4. In consequence of these impulses, the Dominion Parliament embarked in 1879 on the policy of protection. Although the tariff was popularly known as a 20 per cent. tariff, its actual amount was much more. Under the National Policy the average rate of duty on dutiable goods was advanced to 26·11 per cent. in 1880 and to 26·40 per cent. in 1882, falling to 25·32 per cent. in 1883.† The average rate during the tariff for revenue period (1874-78) was 20·45 per cent.‡ The characteristic feature of the tariff of 1879, apart from its being protective, was the abandonment of a large number of specific and the adoption of *ad valorem* duties.

5. In the tariff of 1887 protection was intensified and the dissatisfaction of Nova Scotia with this

---

* Message from the Governor-General, Sessional Papers, 1879, No. 155.

† Maclean, op. cit., p. 27.   ‡ *Ib.*

reversal of the tariff policy to which she had been accustomed, met by an increase in the iron duties. In the tariff of 1887 a large number of *ad valorem* duties were abolished and their places taken either by specific or by compound duties.

6. The next important remodelling of the tariff took place in 1890. Then the principal objects of the Government were described in the Budget Speech of that year to be to admit free of duty those raw materials which might facilitate the development of the country and to reduce the duties upon articles which were not manufactured in Canada, and which were on that account not fit subjects for protection. In the following year, 1891, the principal change was the removal of the duty upon raw sugar and the granting of a bounty of $1 per cwt. upon beet root sugar " produced in Canada wholly from beets produced therein."

7. The fixation of the tariff for 1894 was preceded by an inquiry made by a commission which visited different industrial centres and reported to the Government, and was determined, no doubt, very largely by the contemporaneous discussions on the United States tariff. The characteristic of it was the effort made by the Government to convert specific into *ad valorem* duties and the extent to which this effort was foiled during the passage of the tariff resolutions through committee.

8. The tariff of 1897 is matter of recent history ; the chief points of interest are in the clauses by which a preferential duty is given to the United Kingdom. The action of the Canadian Government has led to notice being given to Germany and Belgium of the desire of England to revise the commercial treaties with these countries.* The question of preferential trade with England had frequently been opened by Canada ; but

---

* Treaties, Belgium, 1862 ; Germany, 1865, both prior to Canadian Confederation.

the reluctance of English statesmen to denounce the treaties always stood in the way of establishing a "preference." It is too soon to determine the precise effect of the abrogation of the treaties upon the commercial relations whether of England and the Continent, or of England and Canada. There seems no reason to doubt, however, as well as much reason to hope, that the general rapprochement between England and her colonies will result in increasing popularity in England of colonial produce, and in increasing, though judicious investment of English capital in colonial enterprises.

TABLE OF THE OCCUPATIONS OF THE PEOPLE OF CANADA, AS DISCLOSED IN THE CENSUS RETURNS, 1890-91.

Productive Classes :

Farmers	408,738
Farmers' sons	240,768
Farm labourers	76,839
Dairymen and dairywomen	907
Apiarists	178
Garden and nursery labourers	2,848
Gardners, florists, nurserymen and vine growers	3,094
Stock-herders and drovers	732
Stock-raisers	958
Other agricultural pursuits	145
Total persons engaged in agriculture	735,207
Miners	13,417
Officials of mining and quarrying companies	242
Quarrymen	1,509
Total persons engaged in mining	15,168
Fishermen	27,079
Lumbermen and raftsmen	12,319
Wood choppers	437
Total persons engaged in lumbering	12,756
Total persons engaged in the exploitation of raw materials.	790,210
Manufactures, total persons engaged in	320,001
Trade and transportation, total persons engaged in	186,495
Professions, total persons engaged in	63,280
Students preparing for professions	10,867
Total professional class	74,147

Non-productive Class :

Indian chiefs	143
Members of religious orders	9,239
Paupers	16,950
Pensioners	179
Retired persons	15,608
Total non-productive class	42,119

TABLE SHEWING POPULATION OF CANADA BY CENSUS AT DIFFERENT PERIODS.

	Nova Scotia.	New Bruns-wick.	Prince Edward Island.	Lower Canada, Quebec.	Upper Canada, Ontario.	Manitoba.	The Terri-tories.	British Colum-bia.
1762..	a 8,104							
1765..				b 69,810				
1817..	c 81,351							
1824..		d 74,176			e 150,066			
1825..				f 479,288				
1827..	g 123,630							
1841..			h 47,042					
1842..					i 487,053			
1844..				j 697,084				
1845..								
1848..			k 62,678		l 725,879	Assiniboia. m 5,391		
1849..								
1851-2	n 276,854	o 193,800		q 890,261	r 952,004			
1856..							s 6,691	
1857..								
1860-1	t 330,857	u 252,047	v 80,857	w 1,111,566	x 1,396,091			
1869..								y 10,586
1870-1	(a)387,800	(a)285,594	(a)94,021	(a)1,191,516	(a)1,620,851	(a)18,995		
1880-1	(b)440,572	(b)321,233	(b)108,891	(b)1,359,207	(b)1,923,228	(b)65,954	(b)56,446	(b)49,459
1890-1	(c)450,396	(c)321,263	(c)109,078	(c)1,488,535	(c)2,114,321	(c)152,506	(c)98,967	(c)98,173

Census of Canada, 1661 to 1871, Vol. IV., Ottawa, 1876—a, p. 61 ; b, p. 65 ; c, p. 82 ; d, p. 84 ; e, p. 83 ; f, p. 88 ; g, p. 94 ; h, p. 132 ; i, p. 134 ; j, p. 144 ; k, p. 174 ; l, p. 164 ; m, p. 175 ; n, p. 232 ; o, p. 224 ; q, p. 202 ; r, p. 178 ; s, p. 242 ; t, p. 344 ; u, p. 332 ; v, p. 358 ; w, p. 282 ; x, p. 254 ; y, p. 376.

(a) Census of Canada, 1870-71, Ottawa, 1873.
(b) Census of Canada, 1880-81, Ottawa, 1884.
(c) Census of Canada, 1890-91, Ottawa, 1893.

Printed by Rowsell & Hutchison, Toronto.

CANADA. GEOLOGICAL FEATURES

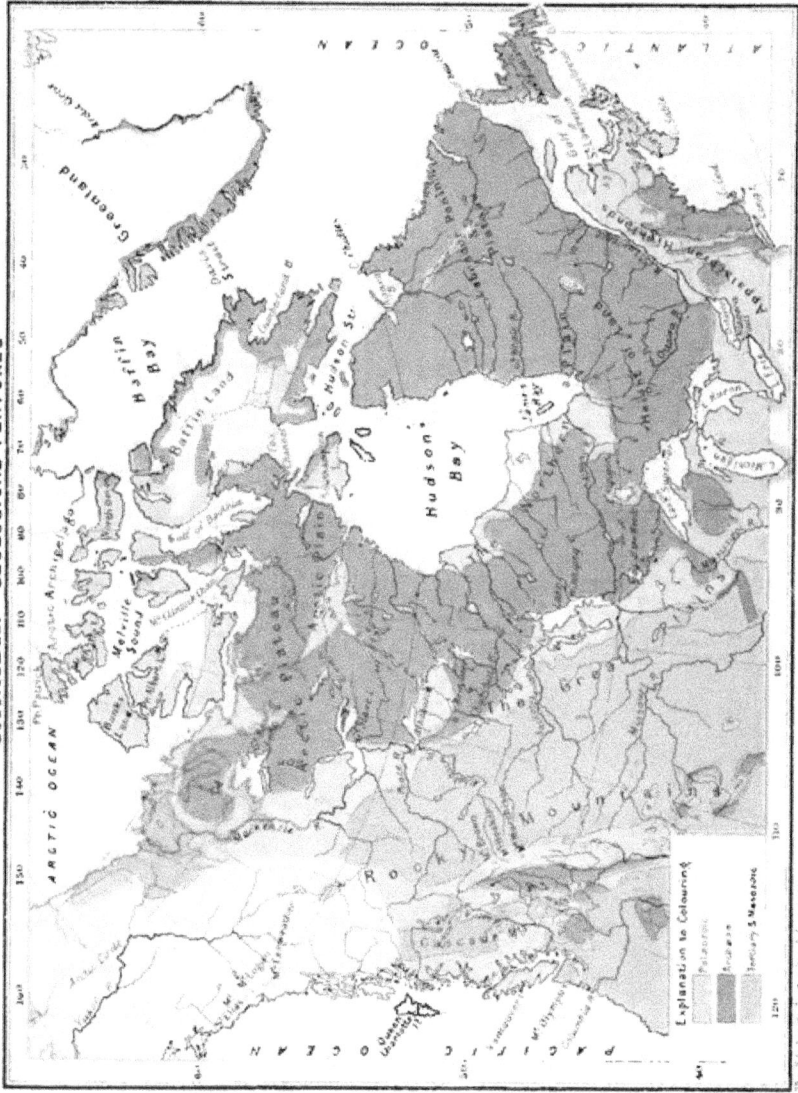

Greenland

Baffin Bay

Baffin Land

Hudson St.

Hudson's Bay

James Bay

ARCTIC OCEAN

Arctic Archipelago

Melville Sound

ATLANTIC OCEAN

PACIFIC OCEAN

Rocky Mountains

The Great Plains

Explanation to Colouring

Palæozoic

Archæan

Tertiary & Mesozoic

British Association Handbook for Canada

CANADA. LAND SURFACE FEATURES

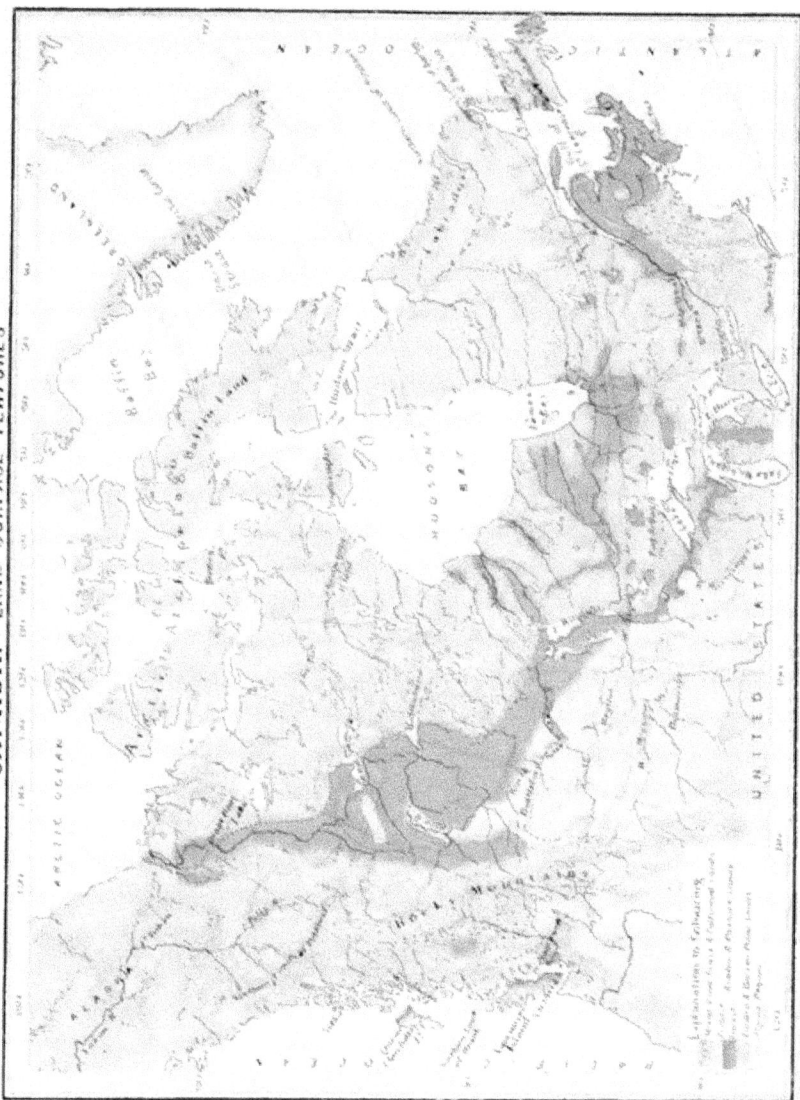

British Association Handbook for Canada

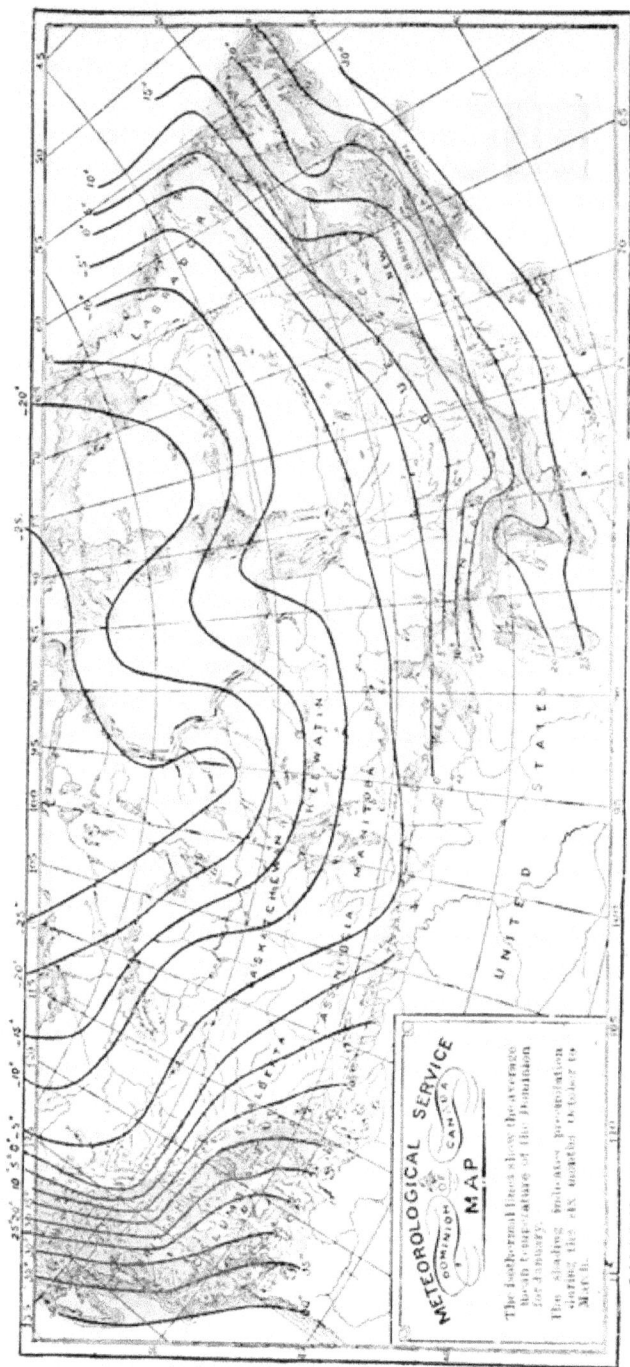

METEOROLOGICAL SERVICE

DOMINION OF CANADA

MAP

The isothermal lines show the average
mean temperature of the Dominion
for January.

The shading indicates precipitation
during the six months October to
March.

1 Under 1 in.   2 Between 12 and 20 in.   3 Over 20 in.

CHART No. 1

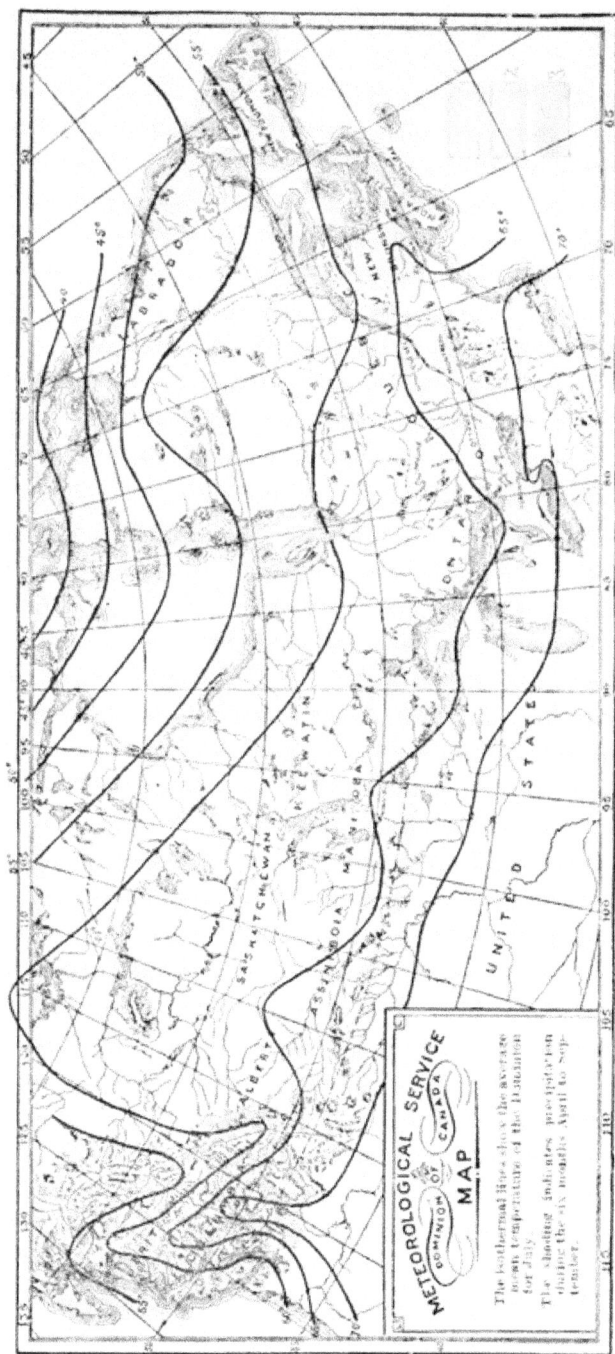

METEOROLOGICAL SERVICE
DOMINION OF CANADA
MAP

The isothermal lines show the average
mean temperature of the Dominion
for July.

The shading indicates precipitation
during the six months April to September.

1 Under ½ in.   2 Between ½ and 2 in.   3 Over 2 in.

CHART No. 2

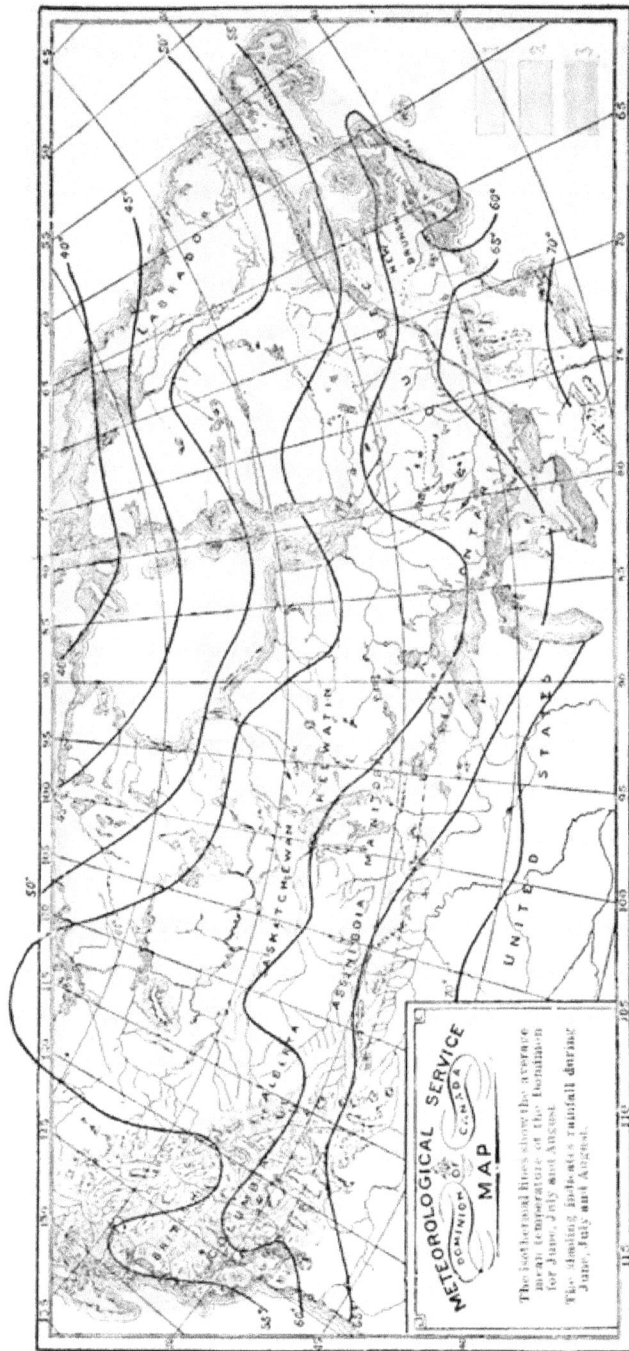

METEOROLOGICAL SERVICE
DOMINION OF CANADA
MAP

The isothermal lines show the average mean temperature of the Dominion for June, July and August.

The shading indicates rainfall during June, July and August.

1 Under 5 in.   2 Between 5 and 10 in.   3 Over 10 in.

CHART No 3

JAN. FEB. MAR. APR. MAY JUNE JULY AUG. SEPT. OCT. NOV. DEC.

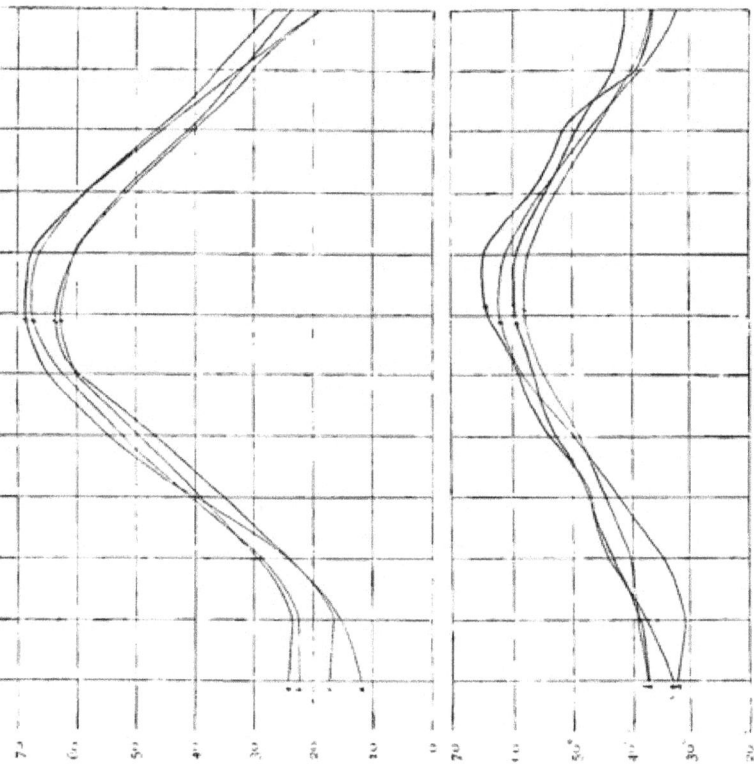

Mean Annual Temperature

1 St. Petersburg (Russia), 38.7
2 Montreal (Canada), 41.8
3 Toronto (Canada), 44.2
4 Christiania (Norway), 41.6

Mean Annual Temperature

1 Edinburgh (Scotland), 46.4
2 Victoria (Brit. Columbia), 48.9
3 Copenhagen (Denmark), 45.4
4 Agassiz (British Columbia), 48.9

CURVES

SHOWING THE AVERAGE MEAN TEMPERATURE OF CERTAIN PLACES
IN EUROPE AND CANADA.

www.ingramcontent.com/pod-product-compliance
Lightning Source LLC
Chambersburg PA
CBHW021345210326
41599CB00011B/755